Experience Analysis for Industrial Planning

Experience Analysis for Industrial Planning

Jean-Paul Sallenave
University of Sherbrooke,
Quebec

Lexington Books
D.C. Heath and Company
Lexington, Massachusetts
Toronto

Library of Congress Cataloging in Publication Data

Sallenave, Jean Paul, 1943-
Experience analysis for industrial planning.

Includes bibliographical references and index.
1. Corporate planning. I. Title.
HD38.S3137 658.4'01 76-7164
ISBN 0-699-00658-0

Published simultaneously in Canada.

Printed in the United States of America.

International Standard Book Number: 0-669-00658-0

Library of Congress Catalog Card Number: 76-7164

Contents

List of Figures

Foreword

In 1968 the Boston Consulting Group published *Perspectives on Experience* [1]. This book made a considerable impact on business strategists by identifying and describing a common pattern of behavior in the unit costs of a product over a period of time, and the way in which this pattern can be exploited for strategic advantage.

The authors of *Perspectives on Experience* intended to draw the attention of business people and prospective clients to the strategic implications of what they dubbed the *experience phenomenon*. They largely succeeded, and their success did not go unnoticed by business academics.

Since 1968 several studies and articles [2] appeared that bore out the findings of the Boston Consulting Group. The most recent and comprehensive research project was the Profit Impact of Market Strategies (PIMS) study undertaken by the Marketing Science Institute [2]. Nothing, however, has been published which would supply business strategists with the technical tools necessary to perform experience analyses and to draw from them implications for strategic action. In the following pages we attempt to fill this void, at least partially.

Acknowledgments

Grateful acknowledgment is offered to the people who contributed to the improvement of the draft of this book, in particular to Prof. Paul H. Beaudoin and Alain Villeneuve of the University of Sherbrooke, and also to Jean-Claude Préfontaine of Bombardier Ltd.

Special thanks are tendered to Dr. William Brandt, Dr. James Hulbert, and Dr. Abraham Schuchman of Columbia University, New York, for their helpful suggestions and guidance.

Experience Analysis for Industrial Planning

1 Industrial Planning and the Leverage Theory of the Firm

Planning is the process by which an organization determines its future flow of resources in accordance with its strategy.

Some firms do not have an explicit strategy and plan only in response to environmental changes—actual or forecast. Other firms do not plan at all; they "hang on" until the competitors drive them out of the market, or until the market dwindles into extinction.

Planning is not easy (1) when the environment is uncertain, (2) when the stakes are high (e.g., capital-intensive industries), (3) when technology is evolving, (4) when markets are highly mobile, (5) when products are undifferentiated between competitors, or (6) when a company is unsure about its future resources. All these conditions, which make planning difficult, are met in the industrial field. Consider for a moment the case of an integrated chemical manufacturer who produces basic chemicals, industrial chemicals, and synthetic fibers and resins. See Figure 1-1.

Generations of business school students remember the case of such an integrated chemical company—Associated Petroleum[a] or Stone Petroleum Company—in which one individual was assigned the task of planning capacity for a chemical product, and this individual did not know (1) which of two technological manufacturing processes would prevail, (2) what the industrial chemical division would buy (this in turn depended on the consumption by the fibers and resins division), and (3) what the market would buy, because of a high rate of product substitution.

Add to the planner's worries the fact that he was only a minor pawn on the corporate chessboard and that there was a minimum 2-year lag between the time a decision was made to build a plant and the time that the plant could start producing.

No wonder that, toward the end of the case, the unfortunate planner was trying not so much to second-guess the demand or the technology, but to find a graceful way to back out of the job.

At first glance, it would seem that the task of planning in the industrial field could be neatly divided into independent parts:

1. Capacity planning
2. Planning for capital expenditures
3. Personnel planning

---

[a] Intercollegiate case clearing house, index 13G150, Harvard Graduate School of Business.

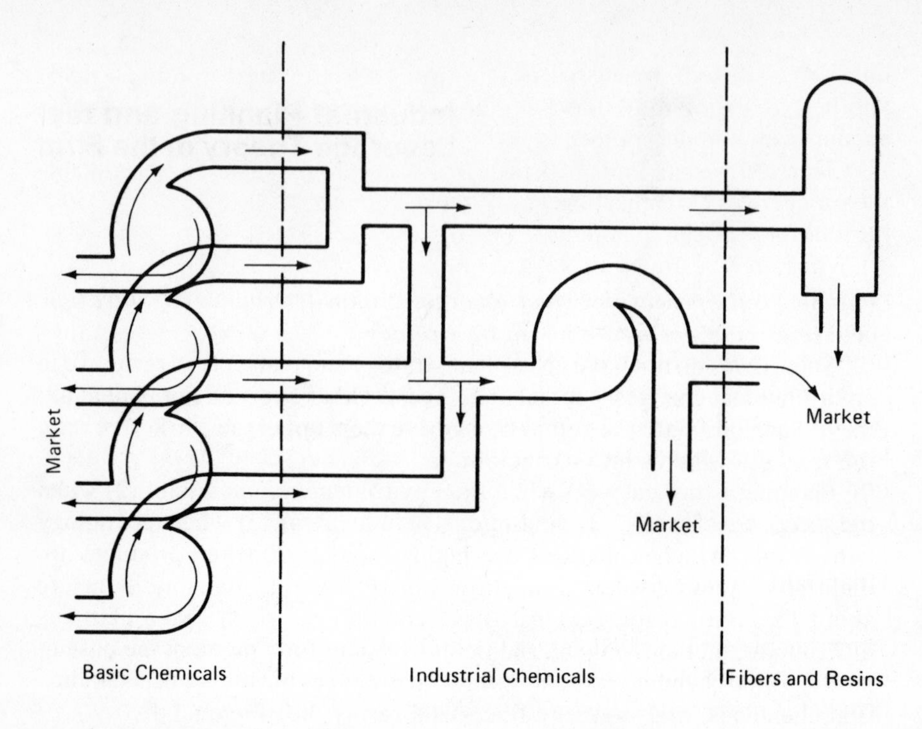

Figure 1-1. Output of an Integrated Chemical Manufacturer.

4. Cost planning and pricing
5. Make or buy decisions
6. Planning "flexible plants" in view of technological changes, etc.

In fact, the worst that could happen to any industrial firm would be to plan the parts independently of one another only to realize, too late, that they would not fit together. *Planning is a global process.* All parts of planning are interrelated. What is needed is a global approach, based on an analysis which deals simultaneously with all aspects of the plan. The *leverage theory* of the firm provides a global understanding of the *resource generation processes* at work in an organization.

The Leverage Theory

A basic function of the enterprise in a capitalist society is to generate funds in order (and not necessarily in *this* order) to pay the stockholders and the employees, and increase the assets of the company. The *resource generation process* is a fundamental attribute of the capitalist firm. Thus the question is raised: What makes one firm generate more resources than another, assuming they start with the same capital? In other words, what accounts for the differences in profitability among firms?

Clearly, if we can find an answer to this question, and if we can control the processes by which a firm generates resources, we shall have a powerful tool for strategic planning.

When an investment of $1 generates a return of $1.50, this process of resource multiplication can be compared with the principle of the lever: a small force applied to one end of the lever results in a greater force at the other end. Similarly, an investment generates a return.

To multiply its resources, i.e., to fund its growth in assets and/or to reward its employees and stockholders, a firm makes use of three possible levers: (1) the financial lever, (2) the marketing lever, and (3) the production lever.

Financial Leverage

The principle of financial leverage is simple: if a firm can borrow money and invest it at a rate higher than the interest rate it is paying for the money, the firm will "make money on money which it does not own."

Example

Company A and Company B contemplate the same investment of $1,000 with an expected return on investment (ROI) of 20 percent. Company A will make the investment with its own capital funds. Company B will fund it half in capital funds and half in borrowed funds (on which B pays 10 percent interest).

	Total Investment	*Equity Invested*	*Debt*	*Cost of Debt*	*ROI*	*ROE*
Company A	$1000	$1000	—	—	20%	20%
Company B	$1000	$500	$500	$50	20%	30%

Company B, using the debt lever, enjoys a better rate of resource generation—relative to its equity investment. Its return on equity (ROE) is

$$\text{ROE} = \frac{\text{profit} - (\text{cost of debt})}{\text{Equity}} = \frac{200 - 50}{500} = 30 \text{ percent}$$

However, if B paid more than 20 percent on its debt, then the debt lever would work against the company, and B's ROE would be smaller than its ROI.

The rate at which a firm is capable of generating resources can be measured; it is expressed by the *sustainable growth rate* of the firm [3]. A company's growth can be generated *internally* by the retention of earnings and their investment in productive assets and/or *externally* by raising new equity or new debt in order to finance productive assets.

The maximum growth in assets that a company can generate internally may be expressed as

$$G_A = \frac{\pi}{E} \cdot (1 - d)$$

where π = profit after tax (PAT)

E = equity

$1 - d$ = retention rate = 1 − dividend rate

Profit can be expressed in a number of ways. Gross profits is

$$\pi_2 = r' \cdot A$$
$$= r'(D + E)$$

where r' = return on assets

A = total assets

D = debt

Profit after interest payments is

$$\pi_1 = r'E + (r' - i)D$$

where i = interest rate on debt. Profit after tax and interest is

$$\pi = (1 - \tau)[r'E + (r' - i)D]$$
$$= (r - i)D + rE$$

where r = net return on assets

τ = tax rate

$$G_A = \frac{\pi}{E} \times (1 - d) \Rightarrow \boxed{G_A = \frac{D}{E}(r - i)(1 - d) + r(1 - d)}$$

G_A is the maximum sustainable growth rate of the company. In the above formula, D is the total debt, E is the stockholders' equity, i is the average interest paid on total debt, and r is defined as (profit after tax + cost of debt)/(total assets).

What can a firm do to improve its rate of resource generation? The

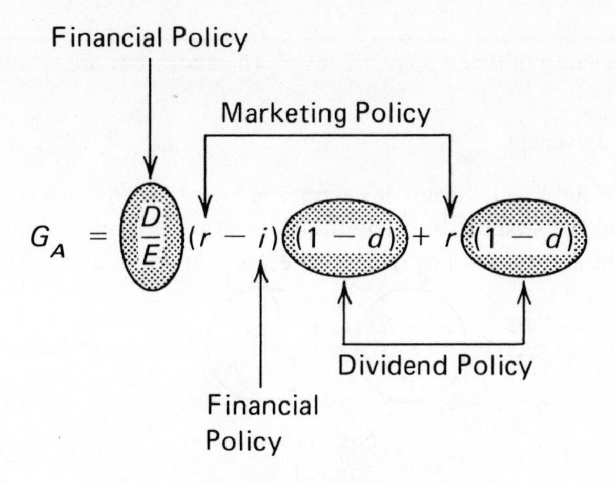

Figure 1-2. Sustainable Growth Formula.

sustainable growth formula gives us all the parameters that a firm can manipulate:

1. It can change its *financial policy* and its capital structure, thus modifying the D/E ratio ($\Delta D/E$).
2. It can improve the profitability of its current investments by producing and selling more goods at a constant price and/or by raising the price. In any event this will entail a decision of *marketing policy* (Δr).
3. It can seek alternate—and cheaper—*sources of financing* to try to reduce its cost of capital (Δi).
4. It can change its *dividend policy* (Δd) so as to distribute less dividends and replow more profits into the company.

See Figure 1-2.

A company will be more or less sensitive to variations of any one of the variables in the growth formula. Typically, we have found that the most sensitive variable is r: that is, the growth of a firm hinges mostly on its ability to generate a greater return on assets. How? By using a marketing lever and/or a production lever.

Marketing Leverage

All firms seek to allocate their resources where they will generate the greatest return. In capital-intensive industries, $1 invested in research and development or production will typically yield a better return than $1 invested in advertising or distribution. In labor-intensive and service industries, the opposite is generally true. Thus it can be said that some firms are

A firm uses one of three possible levels to generate more resources.

Financial Leverage

The use of debt to leverage investments boosts the return on equity (ROE) and the sustainable growth of the firm, as long as the interest paid on the debt does not exceed the net return on assets (ROA)

$$\text{ROA} = \frac{\pi - iD}{D + E} = \left(\frac{\pi - iD}{p \cdot q}\right) \cdot \left(\frac{pq}{D + E}\right) \leftarrow \text{Asset turnover rate}$$

↑ Net return on sales (ROS)

$$\text{ROE} = \text{ROA} \cdot \left(\frac{D + E}{E} \cdot \frac{\pi}{\pi + iD}\right)$$

↑ Leverage factor

Sustainable growth *in assets*:

$$G_A = (1 - d)\text{ROE}$$

$$G_A = (1 - d)\text{ROA} \times \text{leverage factor}$$

$$G_A = \frac{D}{E}\left(\frac{\pi + iD}{D + E} - i\right)(1 - d) + r(1 - d)$$

Sustainable growth *in sales*:

$$G_S = G_A \cdot \frac{pq}{D + E}$$

where π = net profit before interest payments

p = average net price per unit sold

q = number of units sold

D = debt

E = equity

i = average interest rate paid on debt

d = dividend rate

$1 - d$ = earnings retention rate

Marketing Leverage

The most common form of marketing leverage is *market segmentation*, aimed at increasing the return on sales by improving unit margins. Contractual marketing systems such as franchising increase both the sales and the return on sales by multiplying the sources of income of the firm.

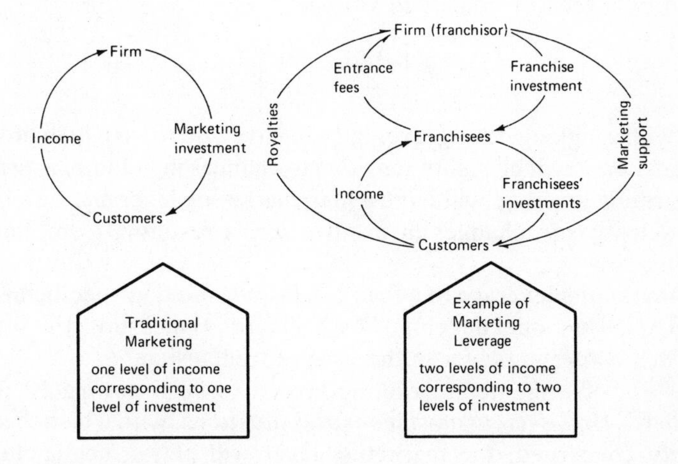

Production Leverage

The accumulation of production of capital-intensive goods results in greater unit profit margins multiplied by a larger volume sold (assuming the accumulated production is actually sold). Production leverage can be measured through *experience analysis*. (See the following chapters).

The rate of average cost decline through the accumulation of production is:

$$\bar{g}_c = 1 - (1 + \gamma)^{-\lambda}$$

The profit growth rate of the firm is:

$$1 + g_\pi = \frac{1 - (1 + \bar{g}_c)(1 - m_t)}{m_t}(1 + \gamma)$$

where γ = production growth rate

λ = elasticity coefficient

m_t = initial average profit margin per unit

Figure 1-3. The Leverage Theory.

more sensitive to a marketing investment and others to a production investment. Those firms with a high marketing leverage will have a low production leverage, and vice versa.

Actually, the concepts of marketing leverage and production leverage can be understood as a consequence of another leverage concept, originated by economists, namely, the operating leverage. Economists define the *operating leverage* of a firm as being the relative change in profits induced by a relative change in volume

$$\left[\frac{\Delta\pi}{\pi} \div \frac{\Delta q}{q}\right]$$

A firm with a high operating leverage will have *ipso facto* a high production leverage; i.e., it will be highly sensitive to changes in volume. A firm with a low operating leverage will have a high marketing leverage; i.e., it will be highly sensitive to changes in profit margins resulting from changes in variable costs.

The marketing leverage of a firm can be measured in specific instances. The FRANSIM model developed by Vielliard [4] measures the sensitivity of a firm's marketing lever in the case of franchising.

The use of a greater marketing lever can be a powerful "resource multiplier." However, in most industrial situations, with which this book is primarily concerned, the marketing lever will play a negligible role in comparison to the production lever, with the possible exception of the marketing leverage derived from licensing agreements and other contractual marketing systems.

Production Leverage

The production lever is the basic resource generation process at work in an industrial situation, denoted by the production of standard industrial goods in a capital-intensive industry: chemicals, minerals, standard electronic equipment, heavy machinery, etc. In those industries planners have always known that the longer the production run, the lower the unit cost. Economies of scale and other factors combine to create productivity gains, which in turn are reflected in the profit of the firm, its reinvestment capacity, and its sustainable growth in assets. By accumulating the production of an industrial product, a firm is usually able to generate more profit per unit. The concept of production leverage refers to this effect of the accumulation of production on unit costs. This effect can be quite dramatic: in one particular instance a large manufacturer of electronic pocket calculators was able to reduce its unit cost by about 50 percent in less than a year.

Clearly, industrial planning has to be concerned with the measurement of production leverage. Cost declines and the ensuing margin changes must be accurately predicted if the industrial planner is to determine the future flow of resources in the firm in accordance with its strategy. The concept of *experience*, which will be introduced in the next chapter, and the analyses that can be derived from this concept enable the industrial planner to measure production leverage, much in the same way as financial leverage or marketing leverage can be measured.

The leverage theory of the firm helps the planner understand the fundamental resource generation processes of the industrial firm. More importantly perhaps, it enables the planner to measure rates of resource generation resulting from the application of any of the three levers of the firm, alone or in combination. Without such measurements planning can easily become an exercise in wishful thinking. "Without numbers I am blind," said Kepler, the seventeenth-century German astronomer, prefiguring the predicament of today's planners. May the following pages help industrial planners sharpen their vision, bearing in mind this postscript from an anonymous student of Kepler's: ". . . and with numbers I am confused." (See Figure 1-3.)

2 The Experience Phenomenon

It has been observed in many industries and for many products that the average total cost per unit (including marketing, distribution, and R&D) declines as a function of the number of units produced. The Boston Consulting Group alone made over 2000 such observations.

This constant decline can be seen as an extension of the learning curve—which applies only to labor costs—to the total cost of a product.

Thus the following empirical law was formulated:

The average total cost per unit of a product, measured in constant dollars, declines by a constant percentage every time experience doubles.

The "experience" which is referred to in this law is defined as the accumulated volume of production. For example, if a firm manufactures a certain type of engine, and if each engine bears a serial number, the experience of the firm at a given moment is equal to the serial number on the engine being manufactured at that moment.

Experience = accumulated volume of production

The experience phenomenon reflects, at the level of the production unit, the productivity increases that economists study at the aggregate level. To be more exact, national productivity increases reflect the accumulation of experience gains by economic agents. In a "steady state," productivity gains by economic agents are constant.

Some remarks are in order, to avoid misinterpreting the experience phenomenon:

(1) The so-called experience "law" is not a natural law (like Newton's laws), but a statistical law (like the normal curve). Costs go down if someone pushes them down!

(2) We have all heard before that "costs go up all the time," and a glance at today's business will drive the point home: labor costs go up, prices of raw materials increase, energy costs more and more, and so does "social" contribution. They increase less, however, if measured in constant dollars—i.e., if the effect of monetary erosion is discounted.[a] To be

[a] See Appendix Table 1.

theoretically correct, the experience law should apply only to the *value added* to the product, i.e., to the elements of cost controlled by the manufacturer. It is only by extension and by approximation that the experience law applies to the total cost, and it has been verified [1] that the experience effect is stronger (i.e., costs decline faster) when the value added to the product accounts for a large percentage of the total cost of the product.

(3) The cost of value added declines for three reasons as production increases:

(*a*) *Economies of scales* occur as production increases. Fixed costs are divided among a greater number of units, thus diminishing the average cost per unit.

(*b*) *Learning curve*: labor cost per unit goes down as workers become more and more familiar and efficient with the production process. They may also improve the process and find "short-cuts."

(*c*) *The cost of capital* goes down as the size of the firm, and that of its assets, increases. Several studies [5] show that the cost of capital in smaller firms is greater than that of the larger enterprises. This is explained by the fact that small businesses do not have access to the equity market and do not generally obtain low-interest loans from their bankers.

(4) The product whose cost pattern is being observed must be stable throughout the period of observation. This condition is met if one studies commodities or raw materials, such as chemicals—a pound of polyvinylchloride is a pound of polyvinylchloride, regardless of the time of observation. However, if we try to study the cost experience curve of airplanes and if we compare on the same graph the costs of a DC-3 and a DC-10, we violate the condition of stability of the product. Under these circumstances, we should find a stable unit of experience. We could study, for instance, the cost pattern of the mile per passenger and predictably find that the experience phenomenon applies to that *stable unit of experience*.

(5) In practice the conditions of validity are not always respected; systematic errors are introduced by measuring the total cost instead of the value-added cost, and sometimes by measuring the average total cost of a product which is not stable over time [6].

These systematic errors do not affect the validity of the implications drawn from experience analyses if (1) value added is a large component of the unit cost, and (2) the product is undifferentiated during the period of observation. As a consequence, experience analyses work best for heavy industrial goods, synthetics, chemicals—and more generally in capital-intensive industries[b]—under normal economic conditions. Nevertheless, experience analyses can be helpful in order to understand the competitive posture of a company and its strategic options, even in situations which do not satisfy the above-mentioned criteria.

---

[b]The experience effect will be stronger for firms with a high operating leverage, due to their economies of scale.

3 The Experience Function

If the unit cost of a product declines by a constant percentage every time experience doubles, the cost function is hyperbolic, and the cost of the *n*th unit can be expressed as a function of the cost of the first unit:

$$C_n = C_1 n^{-\lambda} \tag{3.1}$$

where C_n = the cost of the *n*th unit

C_1 = the cost of the first unit

n = the experience (= accumulated volume of production)

λ = the elasticity coefficient of the function

See Figure 3-1.

A linear expression of the Equation (3.1) can be obtained by a logarithmic transcription:

$$\log C_n = \log C_1 - \lambda \log n \tag{3.2}$$

See Figure 3-2.

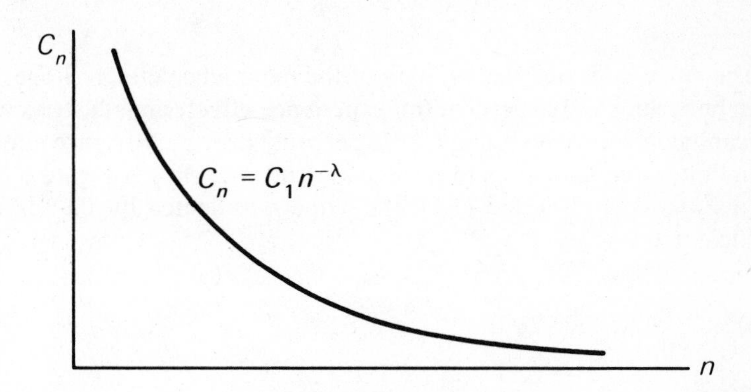

Figure 3-1. Experience Cost Function.

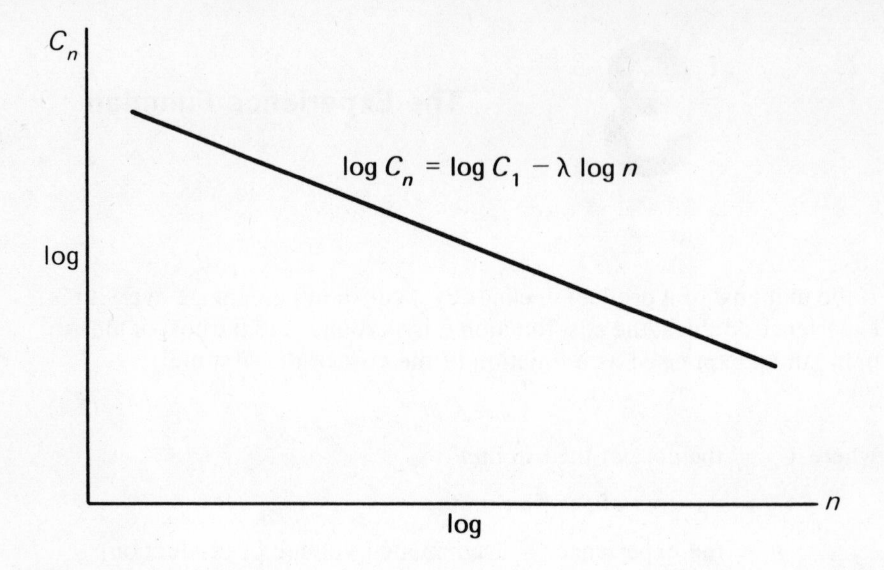

Figure 3-2. A Logarithmic Transcription of the Experience Cost Function.

If the unit cost of a product declines by 20 percent every time experience doubles, the cost of the $(2n)$th unit will be 80 percent of the cost of the nth unit. We shall then say that the *slope of the experience curve* of that product is $k = 80$ percent.

$$k = \frac{C_{2n}}{C_n} \qquad k \leq 1$$

The slope indicates the strength of the experience effect. Slopes may range between $k = 100$ percent (no experience effect), in industries with a low capital intensiveness, and $k = 70$ percent (strong experience effect) in capital-intensive industries (e.g., organic chemistry). See Figure 3-3.

In Equations (3.1) and (3.2), the slope is reflected by the elasticity coefficient λ:

$$k = \frac{C_{2n}}{C_n}$$

$$= \frac{C_1(2n)^{-\lambda}}{C_1 n^{-\lambda}}$$

$$= 2^{-\lambda}$$

$$\log k = -\lambda \log 2 \tag{3.3}$$

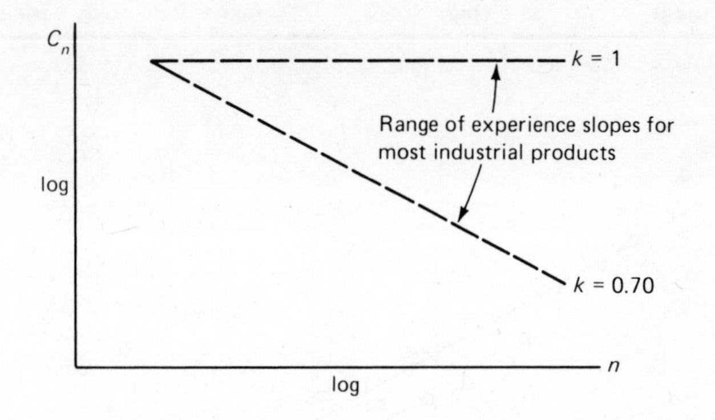

Table of equivalence between $\Delta C(\%)$, k, and λ

Cost decline in % between n and $2n$	$\Delta C =$	0	5	10	15	20	25	30	40	50
Slope of the experience curve	$k =$	1.00	0.95	0.90	0.85	0.80	0.75	0.70	0.60	0.50
Elasticity coefficient of the experience function	$\lambda =$	0^+	0.074	0.152	0.234	0.322	0.415	0.515	0.737	1

Figure 3-3. Table of Correspondence between Cost Decline, Experience Slope, and Elasticity Coefficient.

and

$$\lambda = -\frac{\log k}{\log 2} \tag{3.4}$$

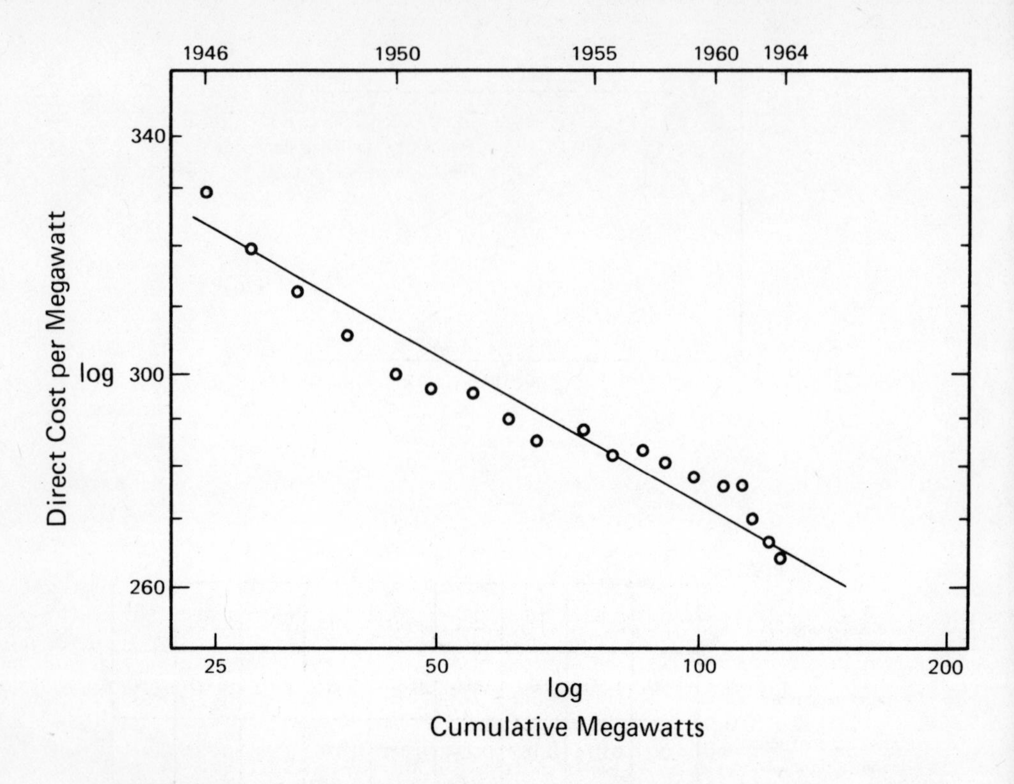

Source: Adapted from the Boston Consulting Group, "Perspectives," (pamphlet). Confidential information from General Electric was made available in public records as the result of antitrust litigation.

Figure 3-4. General Electric: Direct Costs per Megawatt. Steam Turbine Generators (1946-1964).

4 Experience Analyses

Experience analyses apply to cost patterns and to price patterns. They are mostly forecasting techniques. If a firm can predict the cost of a product and its price over a period of time, it will be in a position to foresee its margins, cash flows, and investment capacity. With a minimum of competitive information, the firm can then decide on such major issues as market share objectives, investments and disinvestments, and market segmentation.

Cost Forecasting

Knowing the slope (k) of the experience curve of a product and the rate ($\vec{\rho}$) at which a firm accumulates experience, one can forecast unit costs of the product. See Figure 4-1.

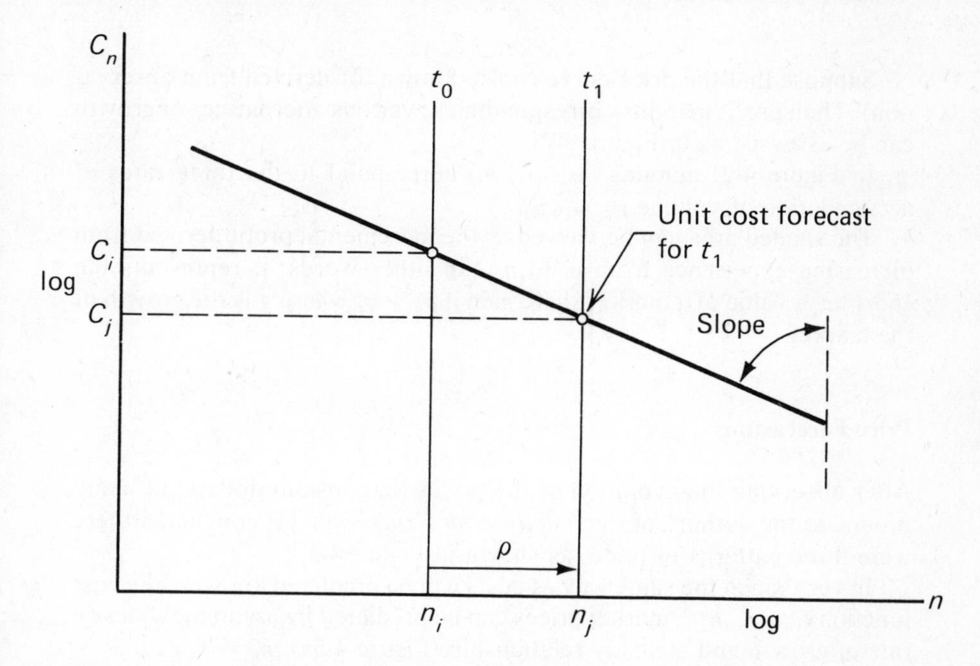

Figure 4-1. Forecasting Units of Production.

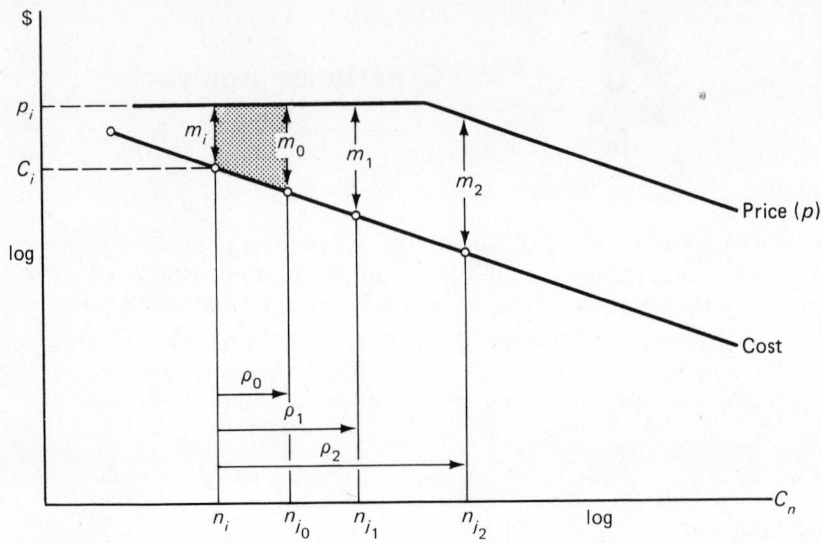

ρ is the *accumulated growth rate* defined as ρ = (production of this year/experience until this year)

Figure 4-2. Profit Margins Corresponding to Various Growth Alternatives Assessed.

Suppose that the price curve is also known (or derived from observation). Then profit margins corresponding to various alternatives of growth can be assessed as in Figure 4-2.

In Figure 4-2, margins m_0, m_1, m_2 correspond to the three rates of accumulation of volume ρ_0, ρ_1, ρ_2.

The shaded area can be viewed as the incremental profit derived from increasing experience from n_i to n_{j0}. In other words, it represents the short-term value of a market share gain if $\rho_0 > g$, where g is the growth of the market.

Price Forecasting

After observing the evolution of the prices (in constant dollars) of many products, the authors of *Perspectives on Experience* [1] concluded there were three patterns of price, as shown in Figure 4-3.

In very much the same way as costs can be predicted knowing the cost function $C_n = C_1 n^{-\lambda}$, market prices can be predicted by assuming a steady rate of growth and a steady relationship (Figure 4-3a): $p_n = p_1 n^{-\lambda}$.

By plotting actual prices of a product, it is possible to discover price

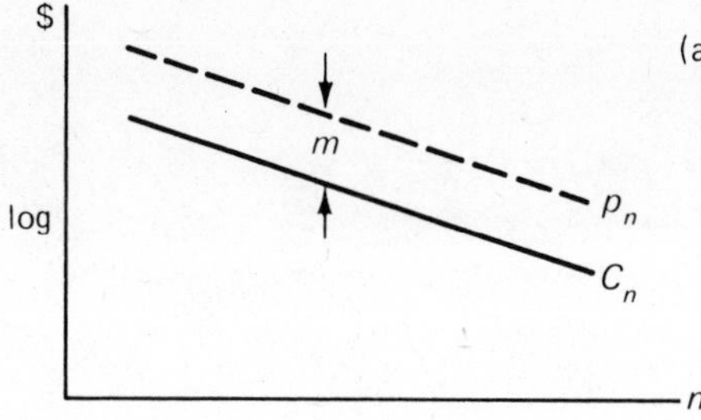

(a) *A stable pattern*

In this pattern, price follows cost and the margin (m) remains constant as time passes and experience accumulates.

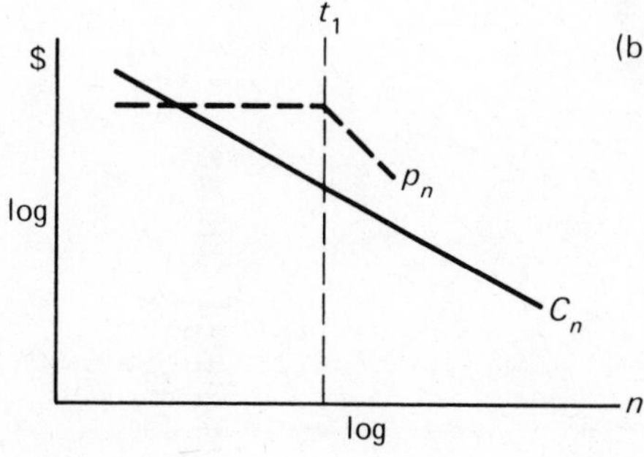

(b) *A stable-unstable pattern*

Until time t_1 producers created a price umbrella—i.e., prices remained constant as cost went down—this obviously attracted more producers and at time t_1 a shake-out erupts: competitors engage in a price war.

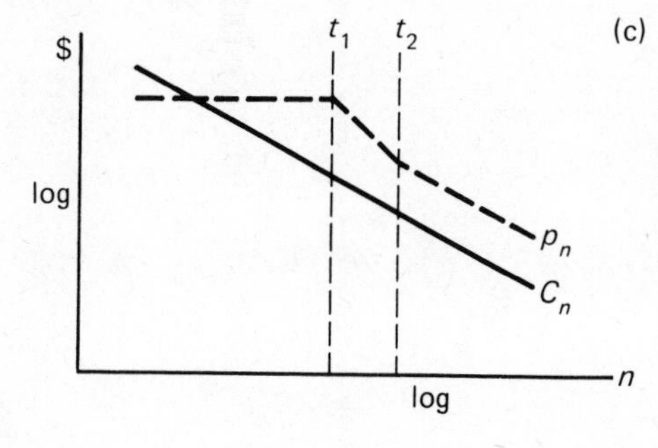

(c) *A stable-unstable-stable pattern*

At t_2 the price war is over. Marginal producers have been eliminated or they have differentiated the product. The surviving producers have more to gain from peaceful coexistence than from fighting it to the end (for what is the end if no one makes a profit?).

Figure 4-3. (a) A Stable Pattern. (b) A Stable-Unstable Pattern. (c) A Stable-Unstable-Stable Pattern.

umbrellas and foresee a price war (Figure 4-3b) especially if the demand for the product is growing, making market share gains through a price war very attractive for the leading producers. (See Figure 4-4).

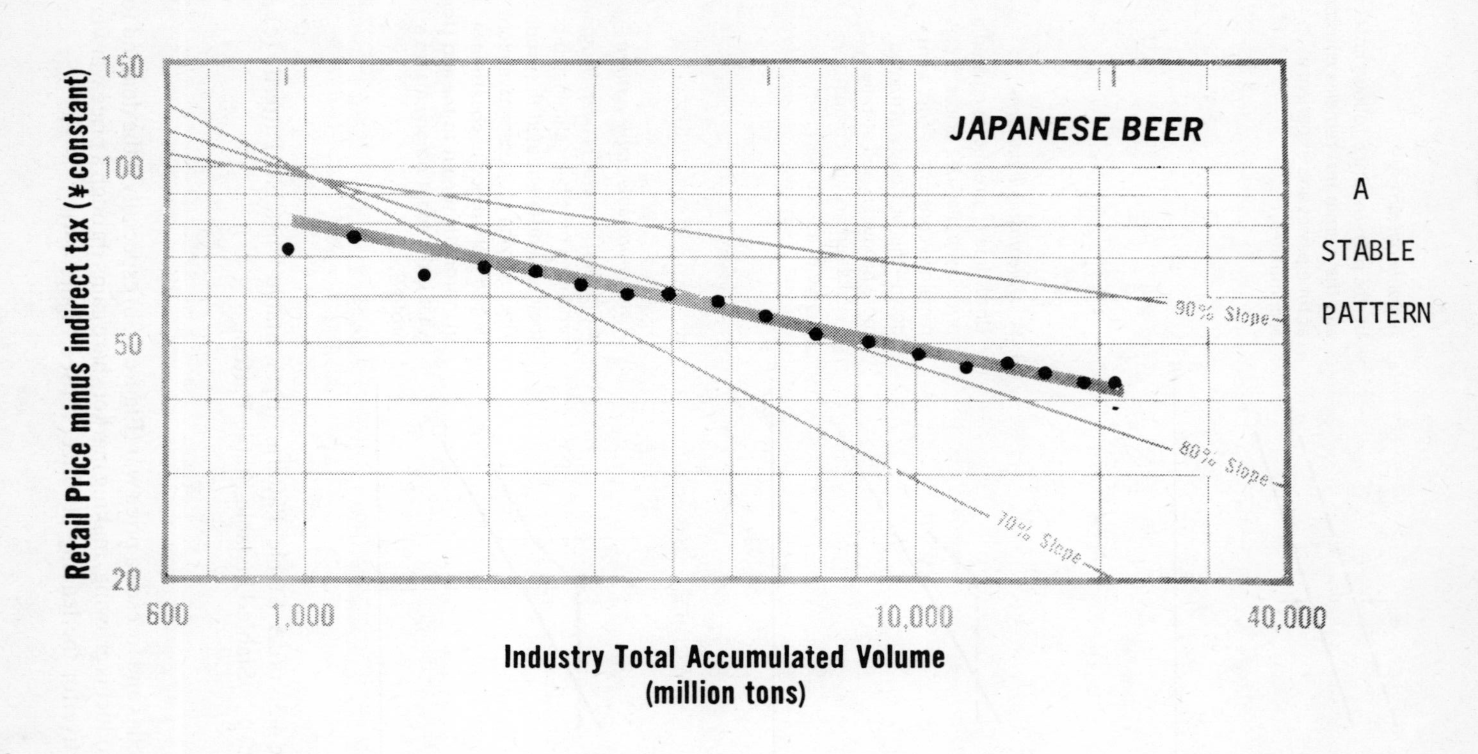

A STABLE PATTERN

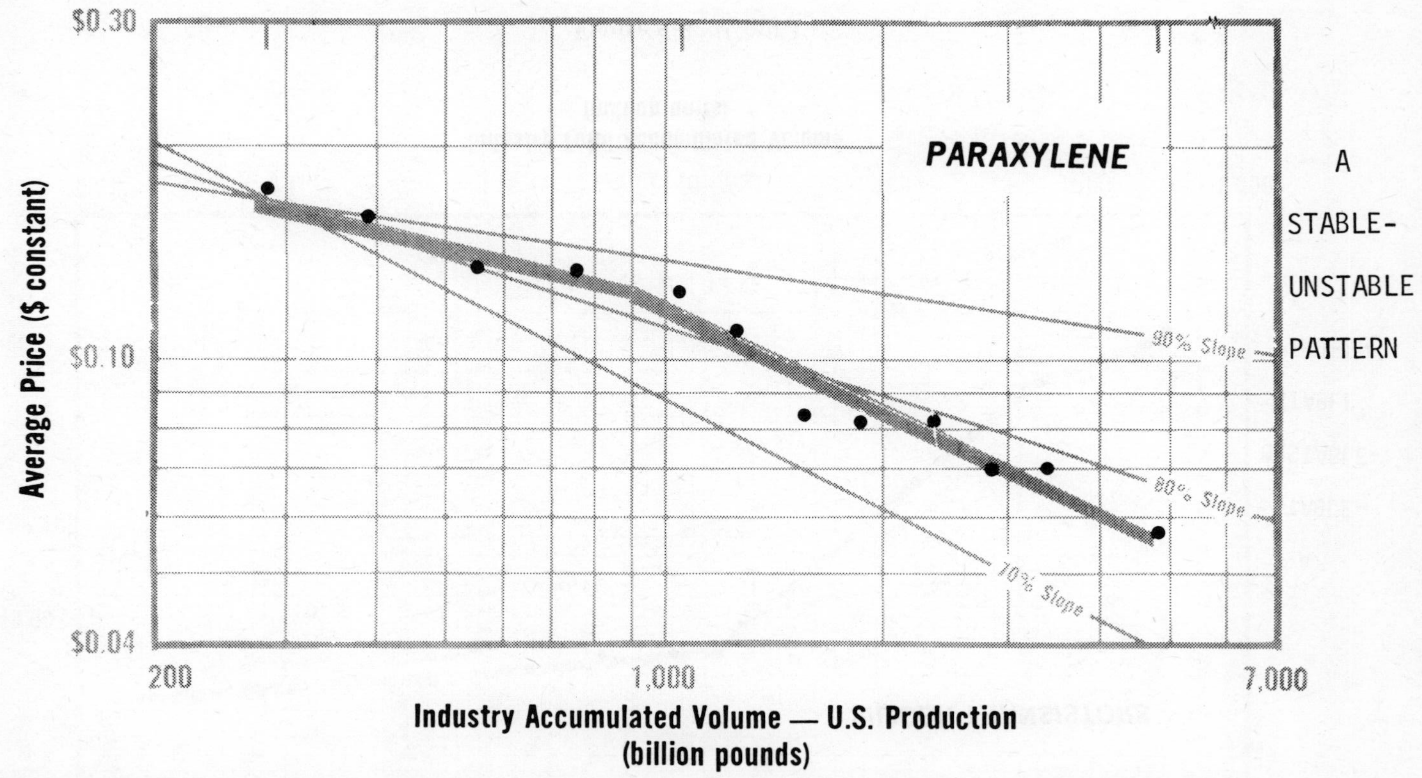

Source: adapted from The Boston Consulting Group; *Perspectives on Experience*, Boston: The Boston Consulting Group, 1968.

Figure 4-4. Price Experience: Three Examples.

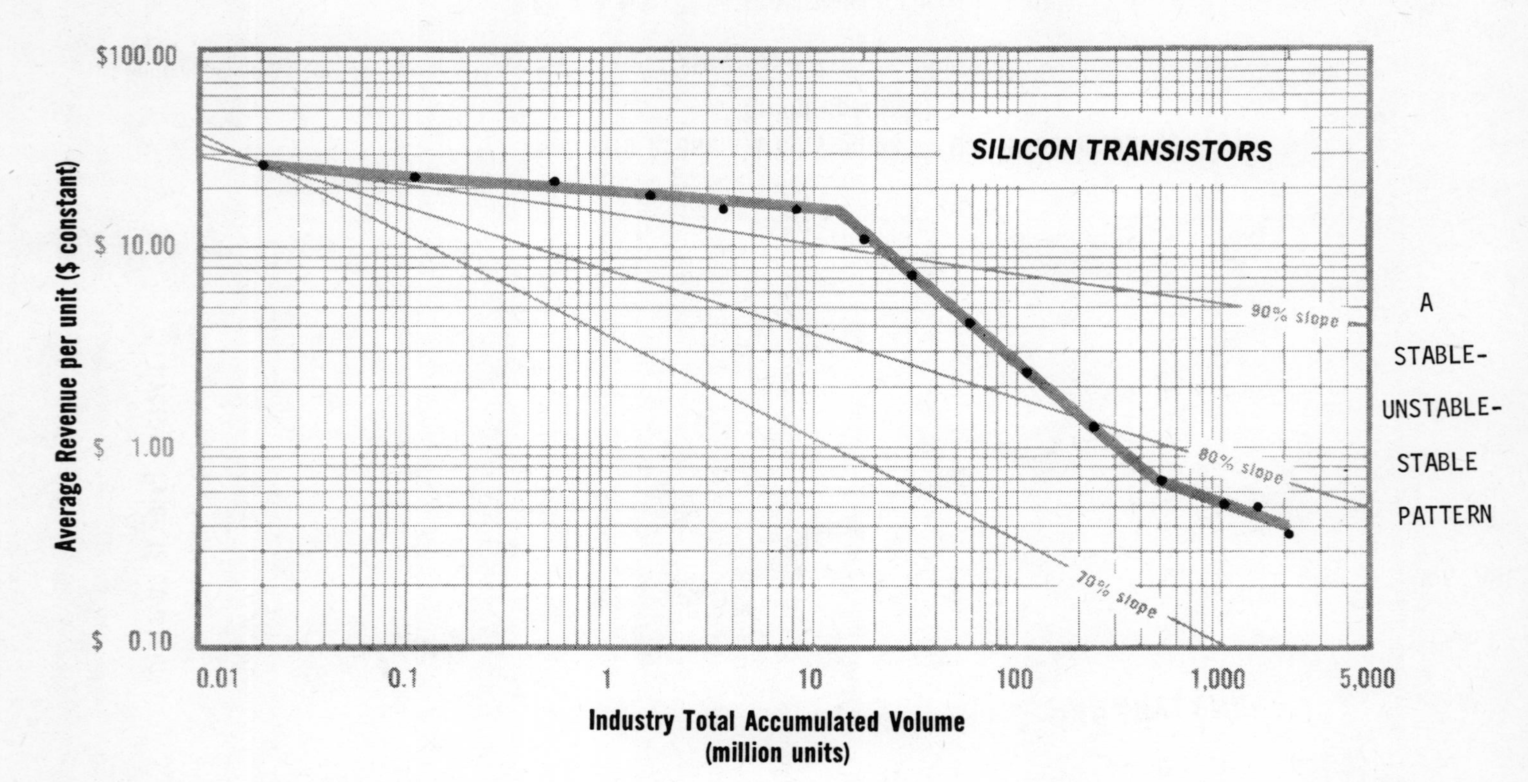

Figure 4-4. (Cont.)

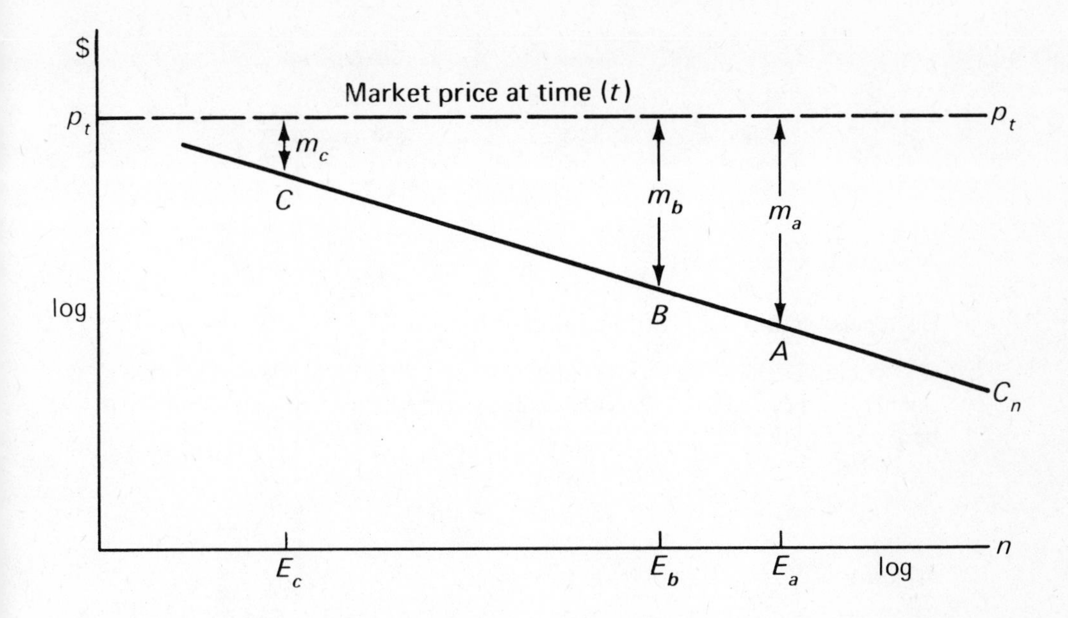

Figure 4-5. Three Different Competitors, A, B, C.

Competitive Planning

Suppose that three competitors A, B, C, have been manufacturing the same product for approximately the same number of years. Further, suppose that they have different market shares. They have different experiences E_a, E_b, E_c.

Any one manufacturer knowing his experience and that of his competitors,[a] and knowing his margin at time t, can approximate his competitors margins at time t. In addition, if A, B, C, move down their experience curve at different speeds, future margins can be assessed to reflect gains or losses in market share. The following textbook [7] example illustrates this point.

Example

A, B, and C are the only manufacturers of product X. They compete in the same market and sell direct; demand grows at 3 percent per annum. The current market price for the undifferentiated product X is $12.

The slope of the cost experience curve of product X is 70 percent.

[a] Or his market share and that of his competitors, since in a steady-state relative market, shares are in the same proportion as relative experiences.

	Unit sales per year	Market share
A	400	58 percent
B	200	28 percent
C	100	14 percent

A's profit margin is $7.

(1) Approximate the profit margins of B and C.

(2) A believes that if he should cut his price by 20 percent, he would drive C out of business and force B to lower his price by 20 percent also. Would this be a good move?

Solution

(1) A's market share is twice that of B. If this relationship has been constant in the past, A's experience is twice that of B. Cost goes down 70 percent when experience doubles; therefore A's cost is 70 percent of B's cost.

A's cost is 12 − 7 = 5. Therefore B's cost is

$$\frac{5 \times 100}{70} \simeq 7$$

C's experience is half that of B; therefore C's cost is

$$\frac{7 \times 100}{70} = 10$$

Profit margins are

A: 12 − 5 = $7

B: 12 − 7 = $5

C: 12 − 2 = $2

Note that B sells twice as many units as C, but his unit margin is more than twice that of C. The experience effect generates a *leverage factor*: it pays less and less (relatively to the competitors) to be the marginal firm.

(2) If A maintained his price at the current level ($12), profits next year for all three competitors would be as follows:

$$\text{A} \quad (7 \times 400) + (7 \times 3\% \times 400) = \$2884$$

$$\text{B} \quad (5 \times 200) + (5 \times 3\% \times 200) = \$1030$$

$$\text{C} \quad (2 \times 100) + (2 \times 3\% \times 100) = \$206$$

Should A cut his price by 20 percent, force C out of business, and split C's share of the market with B proportionately to their respective market shares, then A and B would increase their production:

$$\text{A's production:} \quad 400 + (3\% \times 400) + \frac{2}{3}[100 + (3\% \times 100)] = 481 \text{ units}$$

$$\text{B's production:} \quad 200 + (3\% \times 200) + \frac{1}{3}[100 + (3\% \times 100)] = 240 \text{ units}$$

$$\text{A's profit:} \quad 481 \times (10 - 5) = \$2405$$

$$\text{B's profit:} \quad 240 \times (10 - 7) = \$720$$

If A cuts his price by 20 percent, his profit is reduced by $479, that is, 17 percent, and B's profit dwindles by $310, or 30 percent. Although B is *relatively* more affected than A, A has more to lose than B in money. More importantly, since the growth rate is slow, B's profit cut is not likely to prevent him from satisfying future demand. In other words, the price move envisioned by A is not likely to impair B's *capacity to compete*. If such is the case, A would be wiser not to initiate a price war which would be detrimental to all concerned.

Would the situation be any different if demand grew at 30 percent instead of 3 percent? If demand grows at 30 percent:

$$\text{A's production} \quad P_a = 400 + (30\% \times 400) + \frac{2}{3}[100 + (30\% \times 100)] = 607$$

$$\text{B's production} \quad P_b = 200 + (30\% \times 200) + \frac{1}{3}[100 + (30\% \times 100)] = 303$$

$$\text{A's approximate experience:} \quad \frac{400}{0.30} = 1333$$

$$\text{B's approximate experience:} \quad \frac{200}{0.30} = 667$$

$$C_{n+P_a} = 5\left[1 + \frac{607}{1333}(1.30)\right]^{-0.515}$$

(see the section on Competitive Planning)

$$C_{n+P_b} = 7\left[1 + \frac{303}{667}(1.30)\right]^{-0.515}$$

Appendix Table 3 gives approximate values:

$$C_{n+P_a} = 5\ (0.7890) \simeq \$4$$

$$C_{n+P_b} = 7(0.7890) \simeq \$5.5$$

↑ value read in Table 3

A's profit: $607 \times (10 - 4) = \$3642$

B's profit: $303 \times (10 - 5.5) = \$1364$

B, taking advantage of his lower experience base, partially closes the unit cost gap and increases his profit by 36 percent versus 30 percent for A. It would therefore seem that B stands to gain more than A from the price cut effected by A, assorted with a 30 percent growth. This is, however, misleading insofar as the most relevant question is: Can B finance a necessary production increase of 30 percent with only $1364? If he cannot, then he will not follow the growth of the market, and A will further improve his unit margin relative to B—and he will also significantly reduce B's profit and reinvestment capacity, i.e., *B's capacity to compete in a growing market*.

	Growth Rate	
	3%	30%
Unit cost gap between A and B	$2	$ 1.5
Δ profit from initial situation		
A	−17% ($479)	+30% $842
B	−30% ($310)	+36% $364

Experience analyses revolve around cost forecasting, price forecasting, and competitive planning. Forecasting is always subservient to planning, for in the end only the strategic decision—and the ensuing action—counts. Experience curves can be a powerful tool for strategic planning, yet somewhat unwieldly unless the forecasting techniques derived from the experience analysis are known.

The following chapter will deal with the mathematical aspects of experience forecasting.

5 The Mathematics of Experience

The general "experience equation" expresses the cost of the nth unit as a function of the experience leven n, given the cost of the first unit C_1 and an elasticity coefficient λ:

$$\boxed{C_n = C_1 n^{-\lambda}} \tag{5.1}$$

In practice, C_1 and C_n are rarely known. They are often difficult to obtain from accounting data, since accounting does not provide "total costs" as needed for experience applications, i.e., total costs including R&D, production, marketing, and even service after sale. However, C_n can be deducted from aggregate data, which are readily available, namely from the *accumulated spending* figures.

Accumulated Spending (S_n)

C_n is the cost of the nth unit. As the nth unit is being manufactured, the following sequence of *investments* has already been made:

$$S_n = C_1 + C_2 + C_3 + \ldots + C_{n-1} + C_n$$

where C_n is the cost of the nth unit and S_n is the accumulated spending, at the nth unit.

$$C_n = C_1 n^{-\lambda} \qquad \text{where } \lambda = -\frac{\log k}{\log 2}, \qquad n \in \mathcal{N}$$

S_n can be expressed as a function of n:

$$S_n = \lim \int_{0+\epsilon}^{n} C_1 n^{-\lambda}\, dn$$

$$S_n = \lim_{\varepsilon \to 0^+} \left(\frac{C_1 n^{1-\lambda}}{1-\lambda} \right)_{0+\varepsilon}^{n}$$

$$S_n = \lim_{\varepsilon \to 0^+} \left(\frac{C_1 n^{1-\lambda}}{1-\lambda} - \frac{C_1 \varepsilon^{1-\lambda}}{1-\lambda} \right)$$

accumulated spending formula:

$$\boxed{S_n = \frac{C_1 n^{1-\lambda}}{1-\lambda}} \tag{5.2}$$

or

$$\log S_n = \log\left(\frac{C_1}{1-\lambda}\right) + (1-\lambda)\log n$$

Equation (5.2) can be written as:

$$S_n = \frac{C_1 n \times n^{-\lambda}}{1-\lambda}$$

$$= \frac{n \times C_n}{1-\lambda}$$

This leads us to the relationship between S_n and C_n, as shown in Equation (5.3):

$$\boxed{C_n = (1-\lambda)\frac{S_n}{n}} \qquad (5.3)$$

In most cases, the slope (k) of the experience curve of the product is not known a priori. Therefore λ is not known; however, λ can be derived from the accumulated spending formula. Let

$$S_{n_1} = \frac{C_1 n_1^{1-\lambda}}{1-\lambda} \quad \text{and} \quad S_{n_2} = \frac{C_1 n_2^{1-\lambda}}{1-\lambda}$$

Then

$$\frac{S_{n_2}}{S_{n_1}} = \frac{C_1 n_2^{1-\lambda}/(1-\lambda)}{C_1 n_1^{1-\lambda}/(1-\lambda)} = \left(\frac{n_2}{n_1}\right)^{1-\lambda}$$

$$\log\frac{S_{n_2}}{S_{n_1}} = (1-\lambda)\log\frac{n_2}{n_1}$$

Thus, in Equation (5.4) we have the relationship between λ, S_n, and n:

$$\lambda = 1 - \frac{\log(S_{n_2}/S_{n_1})}{\log(n_2/n_1)} \qquad (5.4)$$

In Chapter 3, λ was written as:

$$\lambda = -\frac{\log k}{\log 2} \qquad (5.5)$$

where k is the slope of the experience curve. Thus Equations (5.4) and (5.5) lead us to

$$\log k = \log 2\left[\frac{\log(S_{n_2}/S_{n_1})}{\log(n_2/n_1)} - 1\right]$$

Let us choose n_2 so that $n_2 = 2n_1$. Then

$$\log k = \log 2\left[\frac{\log(S_{n_2}/S_{n_1}) - \log 2}{\log 2}\right]$$

$$k = \frac{1}{2}\left(\frac{S_{n_2}}{S_{n_1}}\right)$$

The relationship between k and the accumulated spending is shown in Equation (5.6):

$$\boxed{k = \frac{S_{n_2}}{2S_{n_1}}} \quad \text{when } n_2 = 2n_1 \tag{5.6}$$

In practice, experience analyses are always conducted from accumulated spending figures since they are generally available and since they allow the analyst to derive C_n, λ, and k without having to rebuild a complete cost accounting system for the sole purpose of performing an experience analysis.

Since the experience effect is global and continuous, it is always preferable to work from a "global" and continuous basis—that of the accumulated spending of the firm.[a]

Marginal Cost (C_{n+1})

$$\left.\begin{aligned} C_n &= C_1 n^{-\lambda} \\ C_{n+1} &= C_1(n+1)^{-\lambda} \end{aligned}\right\} \frac{C_{n+1}}{C_n} = \left(\frac{n+1}{n}\right)^{-\lambda}$$

Thus, marginal cost, as a function of C_n, is expressed in Equation (5.7):

$$\boxed{C_{n+1} = C_n\left(\frac{n+1}{n}\right)^{-\lambda}} \tag{5.7}$$

The marginal cost equation shows why cost declines, measured in constant $, become smaller and smaller as experience is accumulated:

$$\left(\frac{n+1}{n}\right) \to 1 \quad \text{an } n \text{ becomes large}$$

and

$$C_{n+1} \to C_n$$

[a] However, when the firm is a multiproduct company, the problem of allocating investments to one product or another still has to be resolved.

Future Cost (C_{n+P_i})

Let P_0 be the number of units produced during any given year 0. If the manufacturer increases production every year at a constant rate γ, production during year 1 will be $P_1 = P_0(1 + \gamma)$, and during year i it will be $P_i = P_0(1 + \gamma)^i$.

$$C_{n+P_1} = C_n\left(\frac{n + P_1}{n}\right)^{-\lambda}$$

$$= C_n\left[\frac{n}{n} + \frac{P_0}{n}(1 + \gamma)\right]^{-\lambda}$$

Thus, the cost of next year's last unit will be

$$C_{n+P_1} = C_n\left[1 + \frac{P_0}{n}(1 + \gamma)\right]^{-\lambda} \tag{5.8}$$

At year i the cost of the last unit produced will be

$$\boxed{C_{n+P_i} = C_n\left[1 + \frac{P_0}{n}(1 + \gamma)^i\right]^{-\lambda}} \quad \textit{future cost} \tag{5.9}$$

Equations (5.8) and (5.9) suppose a constant rate of production growth.

Experience tables can be used to find future costs instantly (see Appendix, Table 4).

Future Spending (S_{t+1})

Knowing the investments made this year (S_t) to produce $n_2 - n_1$ units, what investments will have to be made next year (S_{t+1}) to increase the experience from n_2 to n_3, that is, to produce $n_3 - n_2$ units? See Figure 5-1. Assuming a constant growth rate, computations can be made either for a time interval or for an experience interval.

$$S_t = \int_{n_1}^{n_2} C_1 n^{-\lambda}\, dn$$

$$= \left[C_1 \frac{n^{1-\lambda}}{1 - \lambda}\right]_{n_1}^{n_2}$$

$$= \frac{C_1}{1 - \lambda} n_2^{-\lambda} \cdot n_2 - \frac{C_1}{1 - \lambda} n_1^{-\lambda} \cdot n_1$$

$$= \left(\frac{1}{1 - \lambda}\right)(n_2 C_1 n_2^{-\lambda} - n_1 C_1 n_1^{-\lambda})$$

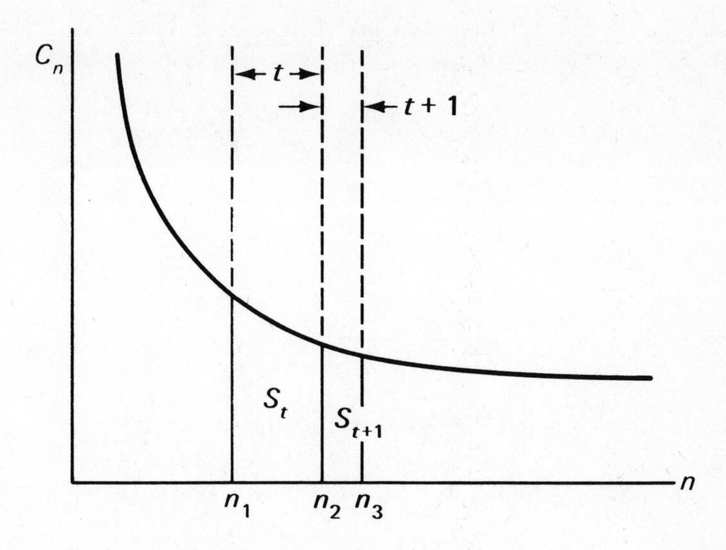

Figure 5-1. Future Spending.

and since

$$C_{n_1} = C_1 n_1^{-\lambda} \quad \text{and} \quad C_{n_2} = C_1 n_2^{-\lambda}$$

this implies

$$S_t = \left(\frac{1}{1-\lambda}\right)(n_2 C_{n_2} - n_1 C_{n_1})$$

Similarly,

$$S_{t+1} = \left(\frac{1}{1-\lambda}\right)(n_3 C_{n_3} - n_2 C_{n_2})$$

$$\frac{S_{t+1}}{S_t} = \frac{[1/(1-\lambda)](n_3 C_{n_3} - n_2 C_{n_2})}{[1/(1-\lambda)](n_2 C_{n_2} - n_1 C_{n_1})} = \frac{n_3 C_{n_3}/(n_2 C_{n_2}) - 1}{1 - n_2 C_{n_1}/(n_2 C_{n_2})}$$

Assuming a constant growth rate γ throughout t and $t + 1$,

$$n_3 = n_2(1 + \gamma)$$

$$n_2 = n_1(1 + \gamma)$$

In addition, C_{n_1}, C_{n_2}, and C_{n_3} can be written respectively as $C_1 n_1^{-\lambda}$, $C_1 n_2^{-\lambda}$, and $C_2 n_3^{-\lambda}$.

Let us rewrite n_2, n_3, C_{n_1}, C_{n_2}, and C_{n_3} in the expression of S_{t+1} as a function of S_t:

$$S_{t+1} = S_t \frac{(1+\gamma)n_2C_1n_3^{-\lambda}/(n_2C_1n_2^{-\lambda}) - 1}{1 - n_1C_1n_1^{-\lambda}/[n_1(1+\gamma)C_1n_2^{-\lambda}]}$$

$$= S_t \frac{(1+\gamma)[(1+\gamma)n_2]^{-\lambda}/n_2^{-\lambda} - 1}{1 - [1/(1+\gamma)]\, n_1^{-\lambda}/[(1+\gamma)n_1]^{-\lambda}}$$

$$= S_t \frac{(1+\gamma)^{1-\lambda} - 1}{1 - 1/(1+\gamma)^{1-\lambda}}$$

$$= S_t \frac{(1+\gamma)^{1-\lambda} - 1}{\dfrac{(1+\gamma)^{1-\lambda} - 1}{(1+\gamma)^{1-\lambda}}}$$

$$= S_t[(1+\gamma)^{1-\lambda} - 1] \times \left[\frac{(1+\gamma)^{1-\lambda}}{(1+\gamma)^{1-\lambda} - 1}\right]$$

Thus next year's spending as a function of this year's spending will be:

$$\boxed{S_{t+1} = S_t(1+\gamma)^{1-\lambda}} \qquad (5.10)$$

The value of $(1+\gamma)^{1-\lambda}$ can be read in Appendix Table 6.

Future *Average* **Cost per Unit ($\bar{C}_{t+1}$)**

Next year's *average* unit cost can be obtained by dividing S_{t+1} by the number of units produced during the period $(t+1)$, that is, $n_3 - n_2$:

$$\bar{C}_{t+1} = \frac{S_t(1-\gamma)^{1-\lambda}}{n_3 - n_2}$$

Similarly,

$$\bar{C}_t = \frac{S_t}{n_2 - n_1}$$

Let P_0 be the production during period t:

$$n_2 - n_1 = P_0$$

and

$$n_3 - n_2 = P_0(1+\gamma)$$

$$\bar{C}_{t+1} = \frac{S_t(1+\gamma)^{1-\lambda}}{P_0(1+\gamma)} = \frac{S_t}{P_0}(1+\gamma)^{-\lambda}$$

$$\bar{C}_t = \frac{S_t}{P_0}$$

$$\frac{\bar{C}_{t+1}}{\bar{C}_t} = (1 + \gamma)^{-\lambda}$$

Next year's average cost per unit as a function of this year's average cost per unit will be

$$\boxed{\bar{C}_{t+1} = \bar{C}_t(1 + \gamma)^{-\lambda}} \tag{5.11}$$

See Appendix, Table 7 for values of $(1 + \gamma)^{-\lambda}$.

Rate of *Average* Cost Decline ($\bar{g}_c$)

The rate of change in the average costs from period t to period $t + 1$ is

$$\frac{\bar{C}_t - \bar{C}_{t+1}}{\bar{C}_t}$$

or

$$\bar{g}_c = 1 - \frac{\bar{C}_{t+1}}{\bar{C}_t}$$

where $\bar{C}_{t+1} = \bar{C}_t(1 + \gamma)^{-\lambda}$. So, the rate of average cost decline is

$$\boxed{\bar{g}_c = 1 - (1 + \gamma)^{-\lambda}} \tag{5.12}$$

Profit Growth Rate ($1 + g_\pi$)

If the price (p) of a product remains constant throughout period t and $t + 1$ while its cost declines because of the experience accumulated between t and $t + 1$, the manufacturer's profit margin will improve between t and $t + 1$. See Figure 5-2. The growth rate of profit can be written as:

$$1 + g_\pi = \frac{\bar{\pi}_{t+1}}{\bar{\pi}_t}$$

where $\bar{\pi}_t$ is the firm's average unit profit during period t and $\bar{\pi}_{t+1}$ is the

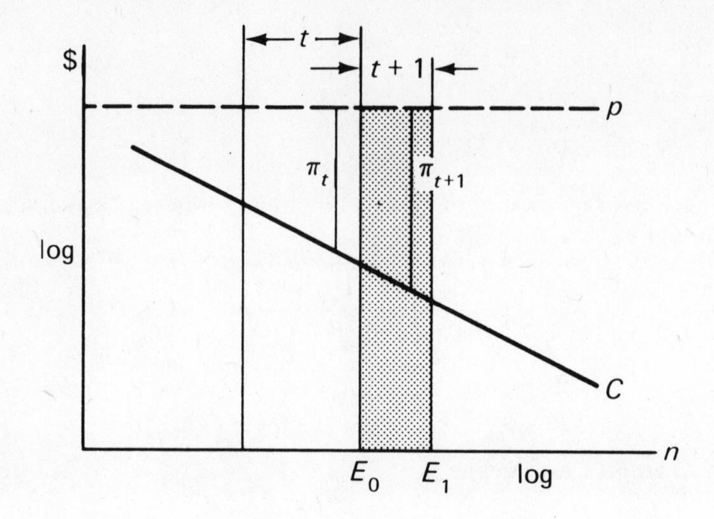

Figure 5-2. Profit Growth Rate.

average unit profit during period $t + 1$. In turn, profit can be written as volume times margin, or

$$\bar{\pi}_t = q_t(p - \bar{C}_t)$$

where q_t is the quantity produced and $\bar{C}_t$ is the average unit cost during period t.

$$1 + g_\pi = \frac{q_{t+1}(p - \bar{C}_{t+1})}{q_t(p - \bar{C}_t)}$$

Let γ be the production growth rate between t and $t + 1$, and $\bar{g}_c$ the rate of average cost change between t and $t + 1$.

$$1 + g_\pi = \frac{q_t(1 + \gamma)[p - (1 + \bar{g}_c)\bar{C}_t]}{q_t(p - \bar{C}_t)}$$

The numerator and the denominator can be divided by p, since p is assumed to be constant between t and $t + 1$:

$$1 + g_\pi = (1 + \gamma)\frac{1 - (1 + \bar{g}_c)\bar{C}_t/p}{1 - \bar{C}_t/p}$$

Let m_t be the initial average profit margin per unit during period t.

$$m_t = \frac{\pi_t}{pq_t} \simeq \frac{q_t(p - \bar{C}_t)}{pq_t} = 1 - \frac{\bar{C}_t}{p}$$

$$\Rightarrow \frac{\bar{C}_t}{p} = 1 - m_t$$

Thus the *profit growth rate* is

$$1 + g_\pi = (1 + \gamma)\frac{1 - (1 + \bar{g}_c)(1 - m_t)}{m_t} \tag{5.13}$$

Equation (5.13) indicates that profit growth is a direct function of the production growth rate [8] and of the rate of cost change. In Equation (5.13) $\bar{g}_c$ is the average cost change. On the other hand,

$$\bar{g}_c = 1 - (1 + \gamma)^{-\lambda} \quad \text{(see Rate of Average Cost Decline)}$$

Therefore

$$\boxed{1 + g_\pi = (1 + \gamma)\frac{1 - (1 + \gamma)^{-\lambda}(1 - m_t)}{m_t}} \tag{5.14}$$

Example

	Company X	*Company Y*
Initial profit margin (m_t)	5%	5%
Sales growth (in volume) (γ)	10%	5%
Slope of the experience curve (k)	70%	70%
Accumulated experience (n)	10,000	10,000
This year's production (P_0)	1000	1000

Demand growth is projected at 5 percent per annum. Company Y plans to follow the growth of the market; Company X plans to increase its market share and to grow at 10 percent, i.e., twice the market growth. For Company Y:

$$1 + g_\pi = 1.05\left[\frac{1 - (1 + \gamma)^{-\lambda}(1 - 0.05)}{0.05}\right]$$

$$= 1.05\left\{1 + [1 - (1.05)^{-0.5146}]\,19\right\}$$

↑ read this value in Appendix Table 7

$$= 1.05[1 + (1 - 0.97521)19]$$

$$= 1.54456$$

For Company X:

$$1 + g_\pi = 1.10\left\{1 + [1 - (1.10)^{-0.5146}]\frac{1 - 0.05}{0.05}\right\}$$

$$= 1.10[1 + (1 - 0.95214)19]$$

$$\uparrow$$

read in Table 7

$$= 2.10027$$

$$\frac{2.10027}{1.54456} = 1.36$$

X will have a 36 percent relative profit growth advantage over Y.

Growth Rate (γ) and Accumulated Growth Rate (ρ)

Let γ be the production growth rate

$$1 + \gamma = \frac{P_1}{P_0}$$

where P_0 is the volume produced during year Y and P_1 is the volume produced during year Y_1.

The accumulated growth rate is the rate of growth of the experience—dn/n—as opposed to the rate of growth of the production $dP/P = \gamma$

$$\rho = \frac{dn}{n}$$

$$\text{where} \quad n = \sum_{j=1}^{T-1} P_j = P_0 + P_1 + P_2 + \ldots + P_{T-1}$$

$$P_1 = P_0(1 + \gamma), \qquad P_2 = P_0(1 + \gamma)^2 \qquad \ldots \qquad P_i = P_0(1 + \gamma)^i$$

$$\sum_{j=1}^{T-1} P_j = P_0[1 + (1 + \gamma) + (1 + \gamma)^2 + \ldots + (1 + \gamma)^{T-1}]$$

$$= \frac{P_0[(1 + \gamma)^{T-1}(1 + \gamma) - 1]}{(1 + \gamma) - 1}$$

$$= P_0\left[\frac{(1 + \gamma)^T - 1}{\gamma}\right]$$

$$\rho = \frac{dn}{n} = \frac{P_T}{\sum_{j=1}^{T-1} P_j} = \frac{P_0(1 + \gamma)^T}{P_0[(1 + \gamma)^T - 1]/\gamma} = \frac{\gamma(1 + \gamma)^T}{(1 + \gamma)^T - 1}$$

$$= \frac{\gamma}{1 - (1 + \gamma)^{-T}}$$

The relationship between the experience growth rate and the production growth rate is

$$\boxed{\rho = \frac{\gamma}{1 - (1 + \gamma)^{-T}}} \qquad (5.15)$$

In Equation (5.15) T is the number of years during which production has been accumulated at rate γ (which is supposed to be constant). If T is sufficiently large, i.e., if the firm has been in business for many years,

$$\rho \simeq \gamma \quad \text{since} \lim_{T \to \infty} (1 + \gamma)^{-T} = 0$$

For all practical purposes, when $T \geq 15$, ρ and γ can be equated in the computations.

Rate of Cost Change

It has already been established that the rate of decline of the average cost between two periods is

$$\bar{g}_c = 1 - (1 + \gamma)^{-\lambda}$$

We shall define the *rate of cost change* as the rate of change in the costs of the last units produced at the end of two consecutive periods.

The rate of cost change is, therefore,

$$g_c = \frac{C_n - C_{n+P_1}}{C_n}$$

$$= 1 - \frac{C_{n+P_1}}{C_n}$$

where $C_{n+P_1} = C_n[1 + (P_0/n)(1 + \gamma)]^{-\lambda}$, as in Equation (5.8).

$$g_c = 1 - \left[1 + \frac{P_0}{n}(1 + \gamma)\right]^{-\lambda} \qquad \text{rate of } marginal \text{ cost change}$$

If the number of units produced is large enough, $C_n \simeq C_{n+1}$ and $C_{n+P_1} \simeq C_{n+P_1+1}$, and g_c expresses the rate of *marginal* cost change.

The rate of *marginal* cost change is therefore a function of three things: (1) the slope of the experience curve (k), or the elasticity coefficient λ($\lambda = -\log k/\log 2$), (2) the production growth rate γ, and (3) the experience growth rate $P_0/n = dn/n = \rho$.

When $\dfrac{P_0}{n} \simeq \gamma$, then $g_c = 1 - (\gamma^2 + \gamma + 1)^{-\lambda}$

whereas

$$\bar{g}_c = 1 - (\gamma + 1)^{-\lambda}$$

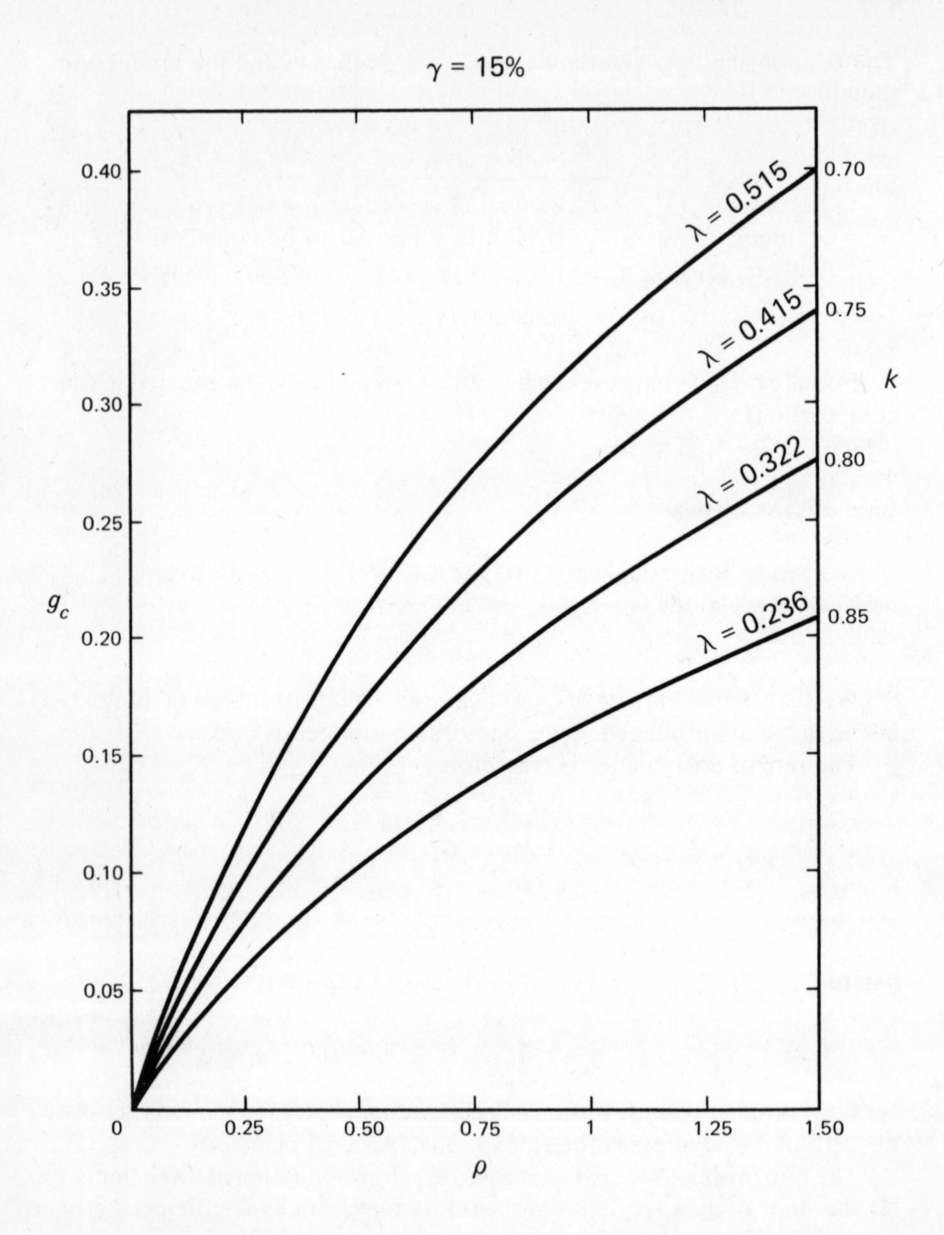

Figure 5-3. Percent Cost Change Corresponding to Different Slopes, with a Production Growth of Fifteen Percent.

An abacus can be constructed that gives the percent cost change corresponding to different slopes k and different experience growth rates ρ, given one production rate γ. In Figure 5-3 the production growth rate of 15 percent has been arbitrarily chosen. One can see immediately that *marginal* cost declines in percent are greater when ρ is large, i.e., when the production is large relative to the experience. This occurs in the growth phase of a new product, which is why it is advantageous for a firm to accumulate experience rapidly during the early years of a product and to gain market share during the growth phase. Larger cost declines in percent mean better unit margins, which will be multiplied by a larger volume (market growth times the growth of the firm's market share).

The rate of *marginal* cost change can also be found in experience tables (Appendix Table 5).

Relative Growth

Although growth per se is important in any industry, since it leads to productivity improvements, i.e., cost declines through the accumulation of experience, the *relative growth* of the companies in the industry is more important in a dynamic and competitive environment. Differences in growth—and therefore differences in *potential profit*—affect the competitive capacity of firms. A firm that grows less than its competitor in a given business, in the long run, will have to divest itself of the business, segment it, or invest heavily to buy up market share and make up for the lost ground.

A company may have an inherent cost advantage over a competitor—for instance, if they operate in different countries with different factor costs and different inflation rates. This inherent cost advantage or disadvantage is *structural*. If a company is at a structural cost disadvantage, it can compensate for it by a *dynamic cost advantage*, i.e., by accumulating experience faster than the competitor who enjoys a structural advantage. In other words, it will have to grow faster than the competitor. How much faster?

The rate of decline of the *marginal* cost is obtained by the derivation of the general cost experience equation:

$$C_n = C_1 n^{-\lambda}$$

$$\frac{dC_n}{C_n} = -\lambda \frac{dn}{n}, \qquad \text{where } \frac{dn}{n} \to \rho$$

(See the section Growth Rate and Accumulated Growth Rate.)

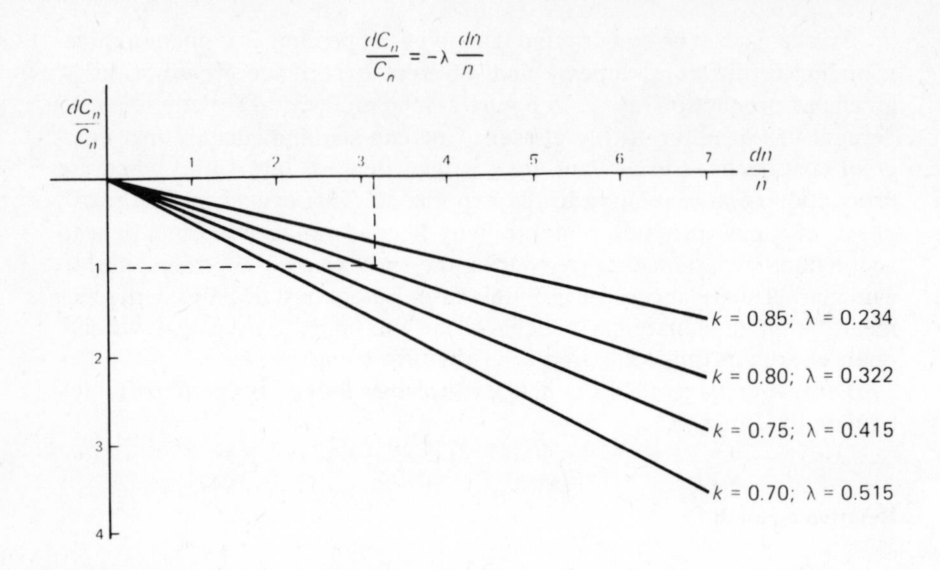

Figure 5-4. Additional Growth Needed to Compensate for a Structural Cost Disadvantage.

A graph can be constructed to read which supplement of accumulated growth will compensate for a 1 percent structural cost differential between two competitors.

It can be seen in Figure 5-4 that 3.1 percent in additional growth[b] is needed to compensate for a 1 percent structural cost disadvantage, if both competitors are on an 80 percent experience curve. This confirms the economists' saying that "1 percent of inflation eats up 3 percent of growth." It is true if the economy as a whole follows an 80 percent experience curve.

Appendix Table 3 gives the additional percentage growth necessary to compensate for a 1 percent cost disadvantage, depending on the cost experience slope (k) of the firm.

The mathematics of experience, as presented in this section, helps the analyst in quantifying the competitive advantage (or disadvantage) of one firm over another. The competitive (dis)advantage can thus be assessed in terms of differentials of cost, growth, profit, and investment capacity (or rate of spending). The only data needed to perform such a *competitive analysis* are aggregate data, since unit cost curves can be derived from the accumulated spending curve (see the section Accumulated Spending S_n).

The following chapter will illustrate the methodology and the usefulness of an experience analysis.

[b] $\rho \to \gamma$, when T is large.

6 Practice: Columbia versus Oxford

The Columbia Corporation and Oxford Ltd. are two fictitious manufacturers of SECs (*s*tandard *e*lectronic *c*omponents). Both companies have a worldwide marketing organization and enjoy a good reputation among their industrial customers, who buy SECs to serve as components in the manufacture of a wide range of electronic equipment and communications systems.

Although the manufacturing process of SECs improved recently with LSI (*l*arge-*s*cale *i*ntegrated circuits) technology, it is still basically a two-phase operation involving the production of "chips" and their assembly on predesigned microcards. Designs are standard throughout the industry, and there can be no difference in the quality of SECs from one manufacturer to another.

Oxford Ltd.

Oxford Ltd. was the first company in the world to develop and manufacture SECs in England for military purposes during World War II. Production ran on a small scale during the next 20 years, and by 1965 Oxford's sales were about half of the sales of their major competitor, the Columbia Corporation. However, Oxford prided itself on the fact that their operations were fully integrated from the production of chips to their testing, screening, and final assembly on microcards. Thus, Oxford Ltd. did not depend on outside suppliers for chips, and did not run the risk of having to cut down production because of a shortage of chips. A worldwide shortage of chips had occurred in the mid-1960s when manufacturers badly underestimated the growth of electronics. During the past decade (1965-1975) the demand for SECs had been growing at rates of 15 to 20 percent per year, in line with the rising demand for electronics—from computers to communications systems. The period 1965-1975 had been a period of growth for Oxford Ltd. Taking advantage of the supply problems of its competitors, the company managed to sextuple its production of SECs in less than 10 years by reinvesting heavily in production capacity.

The success of Oxford Ltd was largely attributed to the aggressive management style of its president Jeremy Cherwell. The company's stock value tripled between 1965 and 1975, and he was knighted in April 1975. Sir Jeremy's aura of success was to be short-lived. Early in 1976 he started to face increasing pressure from the board of directors to improve the profit situation. For years Sir Jeremy had succeeded in convincing the board to agree to a minimal profit in order to capitalize on growth and reap increased profits at a later date. Now the board was pressing for consolidation, not growth, and some members expressed doubts that Sir Jeremy could prove himself to be as good an administrator as he had been an empire builder. Sir Jeremy's task was further complicated by the fact that England was going through its worst economic crisis since World War II with inflation running at 14 percent per year, with no sign of letting up.

Columbia Corp.

In 1956 General Electronics, a large diversified U.S. company involved mostly in the fields of electronics and household appliances, incorporated the Columbia Corporation as a wholly owned subsidiary. Columbia's prime goal was to supply SECs to the parent company; however, it soon became evident that to be profitable, Columbia had to increase production beyond the needs of the parent company. Since the market situation was particularly favorable at the time (Oxford, the major SECs manufacturers in the 1950s had neither the financial means nor the installed capacity to satisfy the current demand), Columbia embarked on a production development program financed by General Electronics. However, progress was hindered by the fact that Columbia had not—and still has not—integrated its production and depended on the parent company for its supply of chips. Columbia merely performed the assembly of the chips on the microcards. In 1963 General Electronics could not supply enough chips for Columbia and forbade Columbia to call on outside suppliers in Japan and Europe, arguing that they did not want to help their competition abroad. Hudson, President of the Columbia Corp., was understandably upset by the situation and engaged in a long battle with the parent management. He fought simultaneously on two fronts: (1) to get rid of the transfer pricing system that had existed since 1956 between General Electronics and Columbia, which, in Hudson's view, resulted in charging Columbia higher than current market prices, and (2) to allow Columbia to have outside suppliers when General Electronics could not supply Columbia with the required components at competitive costs.

After some heated discussions, Hudson was told in polite terms that he was free to leave his post if he did not like the relationship that existed between Columbia and General Electronics. This happened in August 1964. Two months later, General Electronics announced that they could again supply chips to Columbia in sufficient quantity and at a lower price, thus more competitive with what Columbia could have obtained from outside suppliers. In December 1964, Hudson received a substantial salary increase and was confirmed in his post as President of the Columbia Corporation.

In the years following this episode, Hudson avoided all head-to-head confrontation with the parent company. Fortunately, between 1965 and 1975 no major problem arose with the supply of chips, and the Columbia Corporation was able to follow approximately the growth of the SECs market. Although inflation was a source of concern in the mid-1970s, the U.S. economy, with an average annual rate of inflation of 7 percent was still better off than most of the other industrial nations.

Early in 1976 a well-known business magazine published an article on the SECs market which was less than flattering for Columbia. This article depicted Columbia as a slow-moving company who had been lucky to operate in a field where it had not faced much competition until then.

"Battle of the SECs"

The Columbia Corporation, a lesser known subsidiary of General Electronics, has quietly dominated the market for SECs during the past decade. However, the quiet reign of R. Hudson, 60, the inconspicuous President of Columbia seems to be heading towards a period of unrest. The trouble maker is Oxford Ltd. a British company, small by General Electronics standards, yet aggressive and led by Sir Jeremy Cherwell, considered to be one of the brightest stars in the *nouvelle vague* of managers in the City. Interviewed recently, Sir Jeremy declared that in the past ten years his company had experienced tremendous growth without facing much competitive reaction. In a world-market estimated at about 200,000 units, Oxford's sales in 1975 was a comfortable 60,000, a not-so-distant second to Columbia's 80,000. Sir Jeremy further stated that he felt his company had now reached a sales level where it was more important to consolidate the company's market position than to push for more growth, especially in view of the uncertainty of the economy and the galloping inflation in Britain. Most observers however, especially those who have known Sir Jeremy for a long time, disregard his vows of corporate prudence. "Jerry is a tiger, he will go for the jugular until his company is number one" says a long-time friend. Oxford Ltd. and Sir Jeremy may well be the talk-of-the-City, and one would expect Columbia to have some kind of game plan to stall Oxford's advance. Columbia's market share has been slipping badly in the past five years and the net profit is not that impressive either—a mere 10%. But Columbia has always felt secure in the belief that no small company would ever dare to attack General Electronics, even vicariously. When we raised the spectre of Oxford to Mr. Hudson, his answer was a terse: "no comment." In this battle of the SECs no one knows for sure who will eventually triumph: the tiger or the elephant. But it takes more than a "no comment" to impress a stockholder, even though General Electronics is the only stock-holder.

Upon the publication of this article, Hudson was called by the management committee of General Electronics, and he was asked to develop a strategy which would take into account the competitive dimension of the business. This was an indirect criticism of Hudson's previous five-year plans, which were based mostly on a demand analysis and the continuation

Table 6-1
Oxford Limited

Year	*Cash Spent* (Converted in Constant $)	*Production* (Units)
1965	150,000	10,000
1966	162,000	12,000
1967	176,000	14,400
1968	195,000	17,280
1969	220,000	20,736
1970	246,000	24,883
1971	272,000	29,860
1972	312,000	35,832
1973	342,000	41,998
1974	375,000	50,398
1975	415,000	60,478

$1 of income is spent as follows by Oxford Ltd. (average over 5 years):

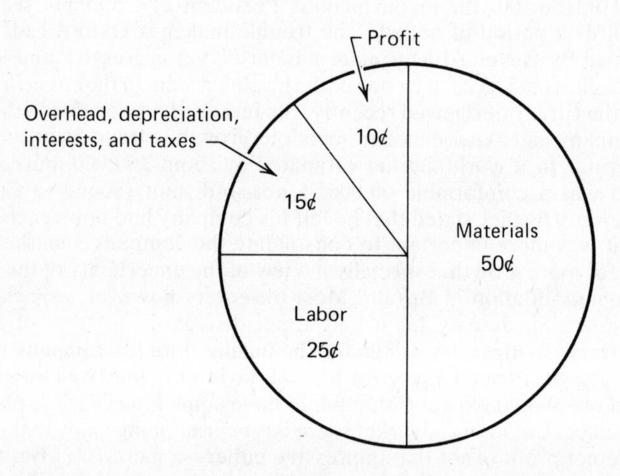

of business as usual. To underscore the importance of Columbia's strategic plan, the management committee told Hudson that they had assigned a management consultant to work with him on this matter. The consultant was to report directly to the management committee.

As a starting point, the consultant recommended that an experience analysis be performed from the data contained in Tables 6-1 and 6-2.

Columbia versus Oxford: Experience Analysis

In this section, a series of questions will be raised which can be answered

Table 6-2
The Columbia Corporation

Year	*Cash Spent* (Current $)	*Cash Spent* (Constant $)	*Production* (Units)
1965	200,000	200,000	20,000
1966	234,330	219,000	23,000
1967	274,776	240,000	26,450
1968	322,186	263,000	30,416
1969	380,131	290,000	34,978
1970	451,622	322,000	40,225
1971	534,260	356,000	46,259
1972	635,890	396,000	53,198
1973	756,002	440,000	61,178
1974	897,168	488,000	70,354
1975	1,062,261	540,000	80,907

$1 of income is spent as follows by Columbia (average over 5 years):

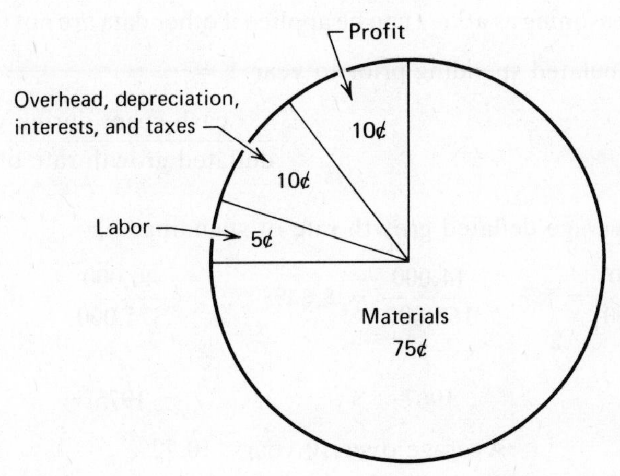

by using the techniques of experience analysis described in Chapter 5. These questions and their answers are not an end in themselves—they merely provide the analyst with a better understanding of the competitive dynamics affecting Columbia and Oxford. The next step consists of putting this understanding to use by plotting a better competitive strategy for Columbia.

What Was Oxford's Experience before 1965?

In the absence of data prior to 1965, the experience can be approximated by assuming that the growth rate of the company was the same in the past.

Oxford's production growth rate between 1965 and 1969 was approximately 20 percent per year, versus Columbia's 15 percent.

$$\text{Previous experience} = \frac{\text{production during 1st year of observation}}{\text{production growth rate}}$$

$$\text{Oxford's experience before 1965} = \frac{10{,}000}{0.20} = 50{,}000 \text{ units}$$

$$\text{Columbia's experience before 1965} = \frac{20{,}000}{0.15} = 133{,}333 \text{ units}$$

What Was the Total Cash Spent by Both Companies prior to 1965?

The same reasoning as above can be applied if other data are not available.

Total accumulated spending prior to year X =

$$\frac{\text{cash spent during year } X}{\text{deflated growth rate of spending}}$$

Oxford's average deflated growth rate of spending:

$$\underset{\substack{\uparrow \\ 1966}}{\frac{12{,}000}{150{,}000}} = 8\%; \quad \underset{\substack{\uparrow \\ 1967}}{\frac{14{,}000}{162{,}000}} = 8.64\% \quad \ldots \quad \underset{\substack{\uparrow \\ 1975}}{\frac{40{,}000}{375{,}000}} = 10.66\%$$

Average over 10 years: 10.72%

Oxford's total accumulated spending prior to 1965:

$$\frac{\text{Cash spent during 1965}}{\text{Average deflated growth rate of spending}} = \frac{150{,}000}{0.1072}$$

$$= \$1{,}390{,}858 \text{ (constant dollars)}$$

Columbia's average deflated growth rate of spending can be computed in the same manner, and it was found to be 10.44 percent.

Columbia's total accumulated spending prior to 1965 is, therefore, 200,000/0.1044 = $1,915,708 (constant dollars).

Table 6-3
Oxford

Year	*Experience*	*Accumulated Spending* (in Constant $)	*Average Cost per Unit* (in Constant $)
Before 1965	50,000	1,390,858	
1965	60,000	1,540,858	15.00
1966	72,000	1,702,858	13.50
1967	86,400	1,878,858	12.22
1968	103,680	2,073,858	11.28
1969	124,416	2,293,858	10.60
1970	149,299	2,539,858	9.88
1971	179,159	2,811,858	9.11
1972	214,991	3,123,858	8.70
1973	256,989	3,465,858	8.14
1974	307,387	3,840,858	7.44
1975	367,865	4,255,858	6.86

Table 6-4
Columbia

Year	*Experience*	*Accumulated Spending* (in Constant $)	*Average Cost per Unit* (in Constant $)
Before 1965	133,333	1,915,708	
1965	153,333	2,115,708	10.00
1966	176,333	2,334,708	9.52
1967	202,783	2,574,708	9.07
1968	233,199	2,837,708	8.64
1969	268,177	3,127,708	8.29
1970	308,402	3,449,708	8.00
1971	354,661	3,805,708	7.69
1972	407,859	4,201,708	7.44
1973	469,037	4,641,708	7.19
1974	539,391	5,129,708	6.93
1975	620,298	5,699,708	6.67

What Were the Average Unit Costs for Both Companies between 1965 and 1975?

$$\text{Average cost in year } X = \frac{\text{cash spend in year } X}{\text{production in year } X}$$

Cost and cash spending figures are in constant $. See Tables 6-3 and 6-4.

What Are the Cost Experience Slopes of Oxford and Columbia?

Let us compute $k = C_{2n}/C_n$ for several values of n (see Tables 6-5, 6-6).

Oxford's average unit cost, measured in constant \$, declines by 25 percent each time experience doubles.

Columbia's average unit cost, measured in constant \$, declines by 18 percent each time experience doubles.

The difference in the cost experience slopes of Columbia and Oxford reflects the differences in the capital intensiveness of their productions. Oxford, being integrated, is more capital-intensive and enjoys a stronger experience effect than Columbia. Oxford has a greater operating leverage

Table 6-5
k Computed for Several Values of n for Oxford

n	$2n$ (approx.)	C_n	C_{2n}	k
60,000	124,416	15.00	10.60	0.71
124,416	256,989	10.60	8.14	0.76
72,000	149,299	13.50	9.88	0.73
86,400	179,159	12.22	9.11	0.75
103,680	214,991	11.28	8.70	0.77
149,299	307,387	9.88	7.44	0.75
179,159	367,865	9.11	6.86	0.75

Average k for Oxford: $k = 0.75$

Table 6-6
k Computed for Several Values of n for Columbia

n	$2n$ (approx.)	C_n	C_{2n}	k
153,333	308,402	10.00	8.00	0.80
308,402	620,298	8.00	6.67	0.83
176,333	354,661	9.52	7.69	0.81
202,783	407,859	9.07	7.44	0.82
233,199	469,037	8.64	7.19	0.83
268,177	539,391	8.29	6.93	0.83

Average k for Columbia: $k = 0.82$

than Columbia. The difference in cost experience slopes may also reflect, in part, a greater production efficiency of Oxford.

What Will Be the Cost of the First Unit Produced in 1976?

$$C_{n+1} = (1 - \lambda)\frac{S_n}{n + 1} \quad \text{See Equation (5.3)}$$

where S_n = accumulated spending until the nth unit and $\lambda = -\log k/\log 2$.
For Oxford:

$$C_{367,865+1} = (1 - 0.415038)\frac{S_{367,865}}{367,865 + 1}$$

$$C_{367,866} = (0.584962)\frac{4,255,858}{367,866}$$

$$= \$6.766 \quad \text{(constant \$)}$$

For Columbia:

$$C_{620,298+1} = (1 - 0.286305)\frac{S_{620,298}}{620,298 + 1}$$

$$C_{620,299} = (0.713695)\frac{5,669,708}{620,299}$$

$$= \$6.512 \quad \text{(constant \$)}$$

Columbia's marginal cost is still lower than Oxford's because of the greater accumulation of experience—in spite of Oxford's stronger experience effect which enables it to diminish its costs every year faster than Columbia.

What Will Each Company Spend in 1976, Assuming the Same Production Growth Rate and Same Rate of Spending as in the Past?

$$S_{t+1} = S_t(1 + \gamma)^{1-\lambda}$$

For Oxford:

$$S_{1975+1} = S_{1975}\underbrace{(1 + \gamma)^{1-\lambda}}$$

$$S_{1976} = 415,000 \text{ (value read in Appendix Table 6)}$$

$$= 415,000(1.11255)$$

$$= 465,700 \quad \text{(constant \$)}$$

For Columbia:

$$S_{1975+1} = S_{1975}(1 + \gamma)^{1-\lambda}$$
$$S_{1976} = 540{,}000\ (1 + 0.15)^{1-0.286305}$$
$$= 540{,}000(1.10484)$$
$$= 596{,}600 \quad \text{(constant \$)}$$

What Will Be the Average *Cost per Unit for Both Companies in 1976?*

$$\bar{C}_{t+1} = \bar{C}_t(1 + \gamma)^{-\lambda}$$

For Oxford:

$$\bar{C}_{1975+1} = \bar{C}_{1975}\underbrace{(1 + \gamma)^{-\lambda}}$$
$$\bar{C}_{1976} = 6.86 \quad \text{(value read in Appendix Table 7)}$$
$$= 6.86(0.92712)$$
$$= \$6.36 \quad \text{(constant \$)}$$

For Columbia:

$$\bar{C}_{1975+1} = \bar{C}_{1975}(1 + \gamma)^{-\lambda}$$
$$\bar{C}_{1976} = 6.67(1 + 0.15)^{-0.286305}$$
$$= 6.67(0.96298)$$
$$= \$6.42 \quad \text{(constant \$)}$$

These results can be verified approximately by dividing the cash to be spent in 1976 by the production forecast:

$$\bar{C}_{1976} = \frac{S_{1976}}{P_{1976}}$$

For example, for Columbia:

$$\bar{C}_{1976} = \frac{596{,}600}{93{,}043}$$
$$= \$6.41$$

(instead of $6.42 found previously)

What Will Be the Marginal *Cost per Unit for Both Companies at the End of 1976?*

This year's marginal costs were (as found above)

Oxford: $6.766

Columbia: $6.512

The cost of next year's last unit—which for all practical purposes can be equated to the marginal cost in 1976—will be:

$$C_{n+P_1} = C_n\left[1 + \frac{P_0}{n}(1 + \gamma)\right]^{-\lambda}$$

For Oxford:

$$C_{n_{1975}+P_{1976}} = C_{n_{1975}}\left[1 + \frac{P_{1975}}{n_{1975}}(1 + \gamma)\right]^{-\lambda}$$

$$C_{n_{1975}+P_{1976}} = 6.766 \left[1 + \frac{60,178}{367,865}(1 + 20)\right]^{-0.413038}$$

↳ read this value in Appendix Table 4

or compute

$$C_{n_{1975}} + P_{1975} = 6.766(1.19728)^{-0.415038}$$

$$\log C_{n+P} = \log 6.766 - 0.415 \log 1.19728$$

$$= 0.83033 - 0.03246$$

$$= 0.79787$$

$$C_{n+P} = \$6.278 \quad \text{(constant \$)}$$

For Columbia:

$$C_{n_{1975}+P_{1976}} = C_{n_{1975}}\left[1 + \frac{P_{1975}}{n_{1975}}(1 + \gamma)\right]^{-\lambda}$$

$$C_{n_{1975}+P_{1976}} = 6.512\left[1 + \frac{80,907}{620,298}(1 + 0.15)\right]^{-0.2803}$$

$$= 6.512(1.1499)^{-0.2863}$$

$$\log C_{n+P} = \log 6.512 - 0.2863 \log 1.14999$$
$$= 0.81371 - 0.01738$$
$$= 0.79533$$
$$C_{n+P} = \$6.242 \quad \text{(constant \$)}$$

What Is the Rate of Average *Cost Decline for Both Companies?*

$$\bar{g}_c = \frac{\bar{C}_{t+1} - \bar{C}_t}{\bar{C}_t} \times 100$$

For Oxford:

$$\bar{g}_c = \frac{\bar{C}_{1976} - \bar{C}_{1975}}{\bar{C}_{1975}}$$

$$\bar{C}_{1975} = 6.86 \quad \bar{C}_{1976} = 6.36$$

$$\bar{g}_c = \frac{6.36 - 6.86}{6.86} \times 100 = -7.28\%$$

or

$$\bar{g}_c = -[1 - ((1 + \gamma)^{-\lambda})] \times 100$$

→ read in Table 7

$$\bar{g}_c = -[1 - (0.92712)] \times 100$$
$$= -7.28\%$$

For Columbia:

$$\bar{g}_c = \frac{\bar{C}_{1976} - \bar{C}_{1975}}{\bar{C}_{1976}}$$

$$\bar{C}_{1975} = 6.67 \quad \bar{C}_{1976} = 6.42$$

$$\bar{g}_c = \frac{6.42 - 6.67}{6.67} \times 100 = -3.72\%$$

or

$$\bar{g}_c = -[1 - ((1 + \gamma)^{-\lambda})] \times 100$$

→ read in Appendix Table 7

$$\bar{g}_c = -(1 - 0.96398) \times 100$$
$$= -3.72\%$$

What Is the Rate of Marginal *Cost Decline for Both Companies?*

$$g_c = \left(\frac{C_{n+P} - C_n}{C_n}\right) \times 100$$

For Oxford:

$$g_c = \frac{C_{n\,1975+P\,1976} - C_{n\,1975}}{C_{n\,1975}} \times 100$$

As computed previously:

$$C_{n\,1975} = 6.766$$

$$C_{n\,1975+P} = 6.278$$

$$g_c = \frac{6.278 - 6.766}{6.766} \times 100$$

$$= -7.21\%$$

For Columbia:

$$g_c = \frac{C_{n\,1975+P\,1976} - C_{n\,1975}}{C_{n\,1975}} \times 100$$

As computed previously:

$$C_{n\,1975} = 6.512$$

$$C_{n\,1975+P\,1976} = 6.242$$

$$g_c = \frac{6.242 - 6.512}{6.512} \times 100$$

$$= -4.14\%$$

What Are the Accumulated Growth Rates of Oxford and Columbia?

The accumulated growth rate is given by

$$\rho = \frac{\gamma}{1 - (1 + \gamma)^{-T}} \quad \text{see Equation (5.15)}$$

Oxford has been in existence since 1946; therefore, we can say that T is sufficiently large so that $\rho \simeq \gamma = 20$ percent for Oxford.

Columbia started its production of SECs in 1956, so for Columbia

$$\rho(1965) = \frac{0.15}{1 - (1.15)^{-8}} = 22.29\%$$

$$\rho(1975) = \frac{0.15}{1 - (1.15)^{-19}} = 16.13\%$$

From year to year the value of ρ approaches that of γ. Thus in 1985 Columbia's accumulated growth rate would be

$$\rho(1985) = \frac{0.15}{1 - (1.15)^{-29}} = 15.2\% \simeq \gamma$$

with the hypothesis of a constant growth rate of $\gamma = 0.15$.

What Will Be the Rates of Growth of Profit in 1976 for Oxford and Columbia, Assuming No Change in Prices?

$$1 + g_\pi = (1 + \gamma)\frac{1 - (1 + \bar{g}_c)(1 - m_t)}{m_t} \quad \text{see Equation (5.13)}$$

For Oxford:

$$1 + g_\pi = (1 + 0.20)\left[1 - 0.9272\left(\frac{1 - 0.30}{0.30}\right)\right]$$

$$= 1.20(1.1698)$$

$$= 1.4037 \rightarrow 40.37\%$$

For Columbia:

$$1 + g_\pi = (1 + 0.15)\left[1 - 0.9628\left(\frac{1 - 0.23}{0.23}\right)\right]$$

$$= 1.15(1.1240)$$

$$= 1.2926 \rightarrow 29.26\%$$

Oxford increases its profit faster than Columbia because (1) it has a smaller experience base (point E_o to the left of the graph in Figure 6-1), and (2) it has a better growth rate $(1 + \gamma)$, multiplied by a faster rate of cost decline g_c.

Note that

$$\left(\frac{\Delta p_{(o)}}{m}\right) \gg \left(\frac{\Delta p_{(c)}}{m'}\right)_c$$

This is somewhat misleading, however, since there is a 7 percent inflation

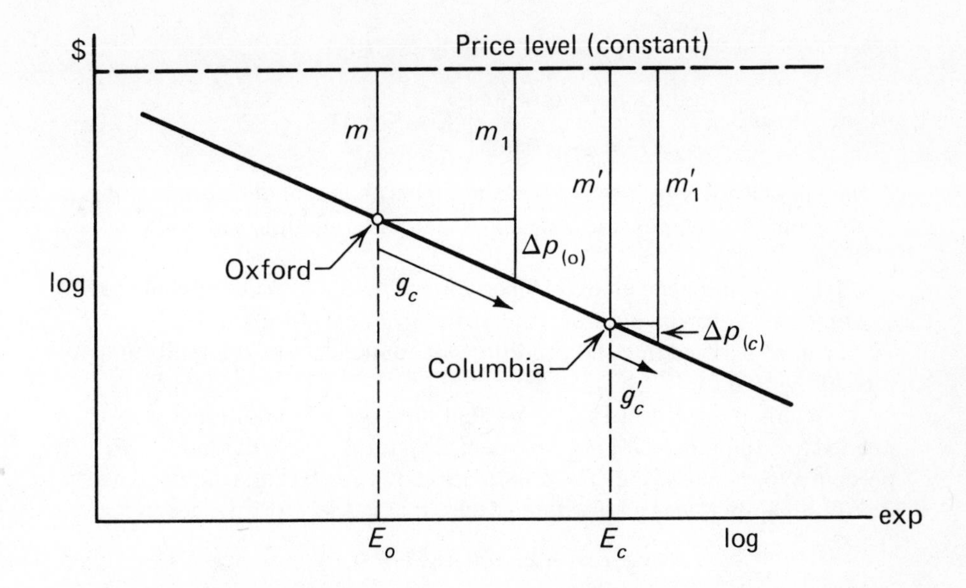

Figure 6-1. Columbia versus Oxford.

gap between Oxford and Columbia; therefore, if prices remain constant in constant dollars, Columbia has a 7 percent price advantage over Oxford. This will no doubt reduce the profit growth gap between the two companies.

What is the Maximum Growth Differential Allowed by the Inflation Gap between Oxford and Columbia?

	Oxford	*Columbia*
Inflation rates	$\zeta_o = 0.14$	$\zeta_c = 0.07$
Elasticity coefficient	$\lambda_o = 0.4151$	$\lambda_c = 0.2863$
Accumulated growth rate	$\rho_o = 0.20$	$\rho_c = 0.1613$

Each manufacturer unit cost is subjected to two forces: (1) inflation (ζ) pushes costs upward, while (2) the experience effect ($\lambda(d_n/n) = \lambda\rho$) drives then downward. The point of competitive *cost* equilibrium is reached when

$$\zeta_o - \lambda_o\rho_o = \zeta_c - \lambda_c\rho_c$$

that is, when

$$\zeta_o - \zeta_c = \lambda_o\rho_o - \lambda_c\rho_c$$

$$\zeta_o - \zeta_c = 0.07 \quad \text{(inflation gap)}$$

$$\left.\begin{aligned} \lambda_o \rho_o &= 0.0830 \\ \lambda_c \rho_c &= 0.0462 \end{aligned}\right\} \Delta = -0.0368$$

Although Oxford's growth is 5 percent higher than that of Columbia, this is insufficient to compensate for the 7 percent inflation gap suffered by Oxford.

Oxford would need an extra 8 percentage points of accumulated growth to compensate for its 3.68 percent *structural cost* disadvantage.

It can be verified that the conditions of competitive *cost* equilibrium are met for $\rho_o = 0.28$.

One can also read in Figure 5-4 that an 8 percent additional growth is needed to compensate for a 3.6 percent structural cost deficiency, with a 75 percent experience slope. The maximum growth differential allowed by the inflation gap between Oxford and Columbia is therefore 12 percent.

$$\Delta\rho = 0.28 - 0.1613 \simeq 0.12$$

In short, Columbia is still in a better strategic posture than Oxford, in 1975, *on a cost basis*. (See Table 6-7). It has a lower average cost per unit and a lower marginal cost. However, Oxford has a smaller experience base and better growth; as a result, Oxford's annual rate of average cost decline is much better than that of Columbia (−7.8 percent versus −3.72 percent), and Oxford's rate of profit growth is also better than Columbia's. In short, from a static standpoint, Columbia is better off than Oxford, but the dynamics of the situation point to Oxford's closing the gap in 1976 and eventually overtaking Columbia if it can maintain its current growth rate and if inflation does not worsen.

Columbia has two weapons to check Oxford's growth. The first weapon is circumstantial: inflation. Inflation erodes Oxford's profits and reinvestment capacity more than it does Columbia's. Columbia's second weapon is its small operating leverage, which makes the company less sensitive to a price change (in absolute dollar terms) than Oxford.

Columbia's loss of market share can be explained only by marketing inefficiency—perhaps by a pricing policy designed to maximize short-term income rather than to defend Columbia's leadership.

Should the growth rate of SECs continue in the future, Columbia would benefit from a more aggressive pricing strategy. Any price cut effected by Columbia would hurt Oxford much more: (1) It would stall Oxford's above-average growth and accumulation of experience, which are its only antidote against inflation. (2) It would cut deeply into Oxford's profit since Oxford has a high operating leverage. (3) It would therefore damage Oxford's reinvestment capacity and its capacity to compete in a growth

Table 6-7
Comparison of Results

		Oxford	*Columbia*
Average unit cost	1975	6.86	6.67
(in constant dollars)	1976	6.36	6.42
Marginal cost	1975	6.766	6.512
(in constant dollars)	1976	6.278	6.242
Rate of *average* cost decline		7.28%	3.72%
Rate of marginal cost decline		7.21%	4.14%
Rate of profit increase		40.37%	29.26%

market; the damage would be compounded by the inflation gap that exists between the two companies.

A close examination of Columbia's marketing policies should follow this experience analysis in order to determine what flexibility Columbia has with respect to pricing.

This may, in turn, lead the consultant to investigate the criteria by which General Electronics assesses its subsidiaries' performance. The conservative pricing policy of a subsidiary could be linked to the performance evaluation system in use by the parent company, if it puts emphasis on contribution margins to the detriment of market share.

7 Conclusion

Experience analyses can provide insight into the competitive dynamics of an industrial business situation. Given the hypotheses of *constant growth*, a *standard product*, and *fixed operating leverage*, which must be made at all times to ensure the validity of the analysis, it is debatable whether experience curves are a good *forecasting tool* insofar as the resulting numbers may depend on conditions of theoretical validity rarely met in reality. However, there is no doubt that experience curves can be extremely helpful as a *planning tool* [8] and as a *learning tool*.

Experience Curves as a Planning Tool

Investment and divestment decisions are all too often made on the basis of simple financial principles such as return on investment. Such principles fail to incorporate two strategic dimensions:

(1) The *business portfolio*. A firm manages a portfolio of products and markets. Planning must have a twofold perspective: planning for each business *(business planning)* and planning the business portfolio *(corporate planning)*.

(2) The *competitive interface*. If competition did not exist, planning would not be necessary and simple forecasting would suffice. Needless to say, anyone in business knows that competition is a fact of life and that financial principles tend to recognize the effects of competition . . . after the fact.

Experience curves are useful tools for business and corporate planning because they offer a visual and conceptual representation both of a firm's business portfolio and of the competitors' portfolios. The competitive interface in each business can be represented on one curve, as shown in Figure 7-1.

The moves of each competitor can be plotted in terms of experience gains, changes in margins, price variations, etc. If more financial information becomes available, then the competitors' capacity to compete can be assessed, as well as their capacity to react to a change in the environment or to a move by another competitor.

By plotting experience curves for each of its businesses, a firm can

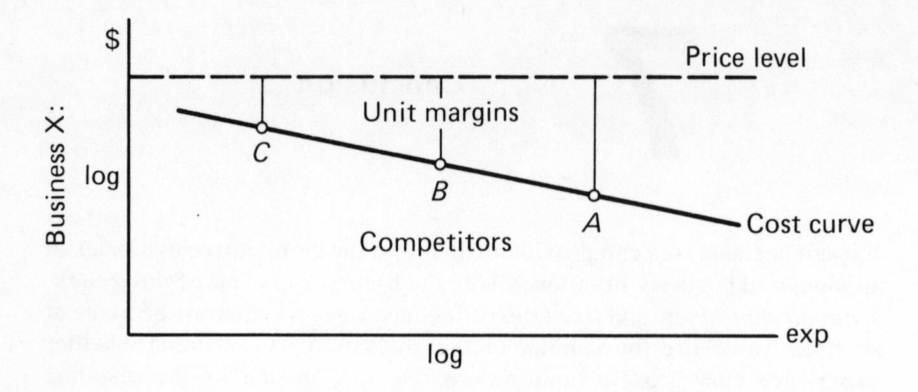

Figure 7-1. Competitive Situation.

assess its relative strength in each business and decide whether to invest more in the business (attack) or to segment it to more manageable dimensions (defense). In any case, experience-curve analyses will help the strategist create a strong business position in a market compatible with the strengths of the firm, rather than to maintain a weak position in a large market. Behind the experience curves lies the belief that a company should strive to dominate its market either by producing more and selling more in the present market, or by producing less (and differently) in a smaller market segment. It is safer to be big in a small market than small in a big market. Strategy is a show of strength. Experience curves provide both a static and a dynamic measure of a company's strength in a competitive business.

Experience Curves as a Learning Tool

Whatever the merits of experience-curve analyses, they cannot be a substitute for strategic thinking. But they can help the strategist develop new insights into the dynamics of business. As such, they are a learning tool which should improve the quality of strategy formulation.

Strategic action implies a commitment of resources, i.e., an investment. Investments can be of two kinds: investments in "fixed" costs (such as *capital expenditures*) or investments in "variable" costs (such as the reduction of unit profit margins). In any case, it should be realized that *cost competition precedes price competition*. Experience analyses, insofar as they provide an insight into the dynamics of costs and prices, are a unique learning tool.

To the functional manager, they will provide a comprehensive frame of reference. He will thus understand his business not in terms of marketing, nor in terms of finance or production, but rather in terms of the ability of the business to survive and thrive in its competitive environment. This is, after all, a much more basic and immediate frame of reference for any business manager.

The experience-curve phenomenon is without doubt the cornerstone of capitalism since it depicts *productivity gains*, which in turn lead to *capital formation*. At the microeconomic level, experience analyses test the economic efficiency of the firm and provide a basis of comparison with competitors. They also test the relevance of past strategies, bearing in mind that strategies are neither good nor bad, only better or worse than those of the competitors. Competition is a fact of life, at the national or international level. It may be seen as a game—and a dangerous one at that—in which experience analyses provide both an insight into the rules of the game and the ultimate score card.

8 Experience Tables

Experience tables enable the analyst to take a short-cut in what would otherwise be lengthy computations.

When plotting a strategy, or merely in the planning phase of a project, the analyst asks "what if" questions—e.g., what if demand grew at 10 percent instead of 5 percent? What effect would it have on cost declines? The tables presented in the Appendix will provide an instant answer to this type of sensitivity analysis. Furthermore, they can be used without in-depth knowledge of experience analysis, although without proper understanding of the technique the analyst runs the risk of overlooking the conditions of validity of the tables.

Table 1: Implicit Price Deflators for U.S. Gross National Product

Experience analyses must be performed in constant currency. The value of $1 erodes year after year. Therefore, today's dollar must be multiplied by a corrective factor to be expressed in, say "1972 dollars." (1972 has been chosen arbitrarily.)

Table 1 lists the corrective factors which, when multiplied by current dollars of any given year (between 1946 and 1975), will give the sum in constant 1972 dollars.

Use of Table 1

Multiply the sums expressed in current dollars by the implicit price deflator of the corresponding year to obtain the sums in constant (1972) dollars. For example,

Year	*Unit cost (current $)*	*Implicit price deflator*	*Unit cost [constant (1972) $]*
1972	89	1.00000	89
1973	92	0.94410	86.86
1974	99	0.86058	85.20
1975	101	0.79145	79.94

read in Table 1 → (implicit price deflator column)

Note in this example that unit costs in constant dollars go down, although they increase if measured in current dollars (as in accounting statements).

Table 2: Equivalence between the Slope and the Elasticity Coefficient

Table 2 gives the value of λ corresponding to a given experience slope k.

Use of Table 2

First, approximate the experience slope. You can plot unit costs as a function of experience and draw a regression line between the points. Read the value of k by drawing a parallel from point [1,10]; it intersects the vertical line [2,1] at k. See Figure 8-1.

Or, you can use the formula

$$k = \frac{C_{n_2}}{2C_{n_1}} \quad \text{when } n_2 = 2n_1$$

See Equation (5.5).
Second, read the value of λ corresponding to the slope k in Table 2.

k		λ
—		—
0.81	→	0.304
—		—

Table 3: Growth Needed to Fill Cost Gap

Column 3 of Table 3 gives the additional percentage growth necessary to compensate for a 1 percent cost disadvantage, depending on the cost experience slope (k) of the producer.

Example

Suppose a domestic manufacturer suffers from a 5 percent structural cost disadvantage in comparison with a foreign competitor. Both are on a 76 percent experience curve. How much faster should the domestic company increase its experience to make up for its structural cost disadvantage vis-à-vis its foreign competitor, *ceteris paribus*?

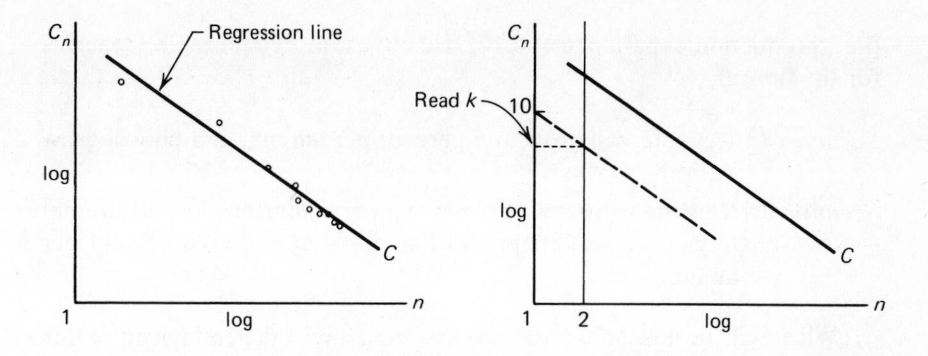

Figure 8-1. Two Steps to Find the Experience Slope (k).

Use of Table 3

First, read the additional percentage growth necessary to compensate for a 1 percent cost disadvantage.

k	λ	$\Delta\rho$
—	—	—
0.76	0.3959	2.5257
—	—	—

Second, multiply the number found in the third column by the percent of cost disadvantage:

$$2.5257 \times 5 = 12.6285$$

Answer:

12.6% ← additional percentage growth.

Table 4: Future Costs

Given an experience slope k, an accumulated growth rate ρ, a projected growth rate γ, and the current cost C of a product, Table 4 allows one to work out what the unit cost will become in the future ($1 < \text{year} < 10$). *ceteris paribus*.

Example

A manufacturer's average unit cost today is \$10. He operates on a 80 percent experience curve, with an accumulated growth rate of 10 percent (i.e., production/experience $\simeq$ 0.10). He envisions two possible scenarios for the future:

Scenario 1: Demand will grow at 5 percent per annum, and he will grow with the market.

Scenario 2: Demand will grow at 10 percent per annum and he will attempt to gain market share and thus grow at about 20 percent per annum.

What will be his unit cost 5 years from now, depending on which scenario materializes?

Use of Table 4

First, go to the page corresponding to the value of ρ. Say that $\rho = 0.10$. Second, select the table with the proper value of k. Say that $k = 0.80$. Now read the multiplying factors corresponding to the horizon (year) and the projected growth rates.

% Growth	5	20
Year		
—		
—		
—		
—		
5	0.8630	0.8143
—		
—		

Finally, multiply by the initial cost:

Scenario 1: Unit cost in 5 years will be: 0.8630 × 10 = \$8.63.

Scenario 2: Unit cost in 5 years will be: 0.8143 × 10 = \$8.14.

Table 5: Rate of Marginal Cost Decline

Given an accumulated growth rate ρ, an experience slope k, and a projected growth rate γ, Table 5 gives the annual rate of cost decline.

Example

Company XYZ has an accumulated growth rate of 50 percent (i.e., production/experience = 0.5) and operates on an 80 percent experience curve. Should it increase its annual production by 10 percent, what will be its rate of unit cost decline between the last unit produced this year and the last unit to be produced next year?

Use of Table 5

Start by looking for the page corresponding to the value of ρ. Suppose ρ = 0.50. Read the rate of cost decline corresponding to k and γ.

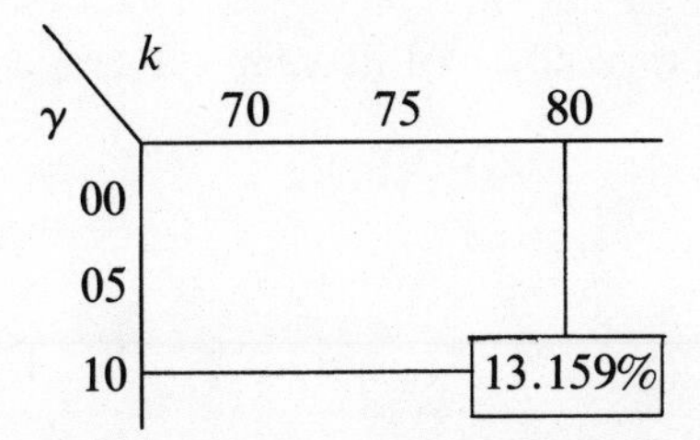

Caution: The value of ρ will change year after year if the experience base is small at the outset; in this case, one should not compound the same rate of cost decline over several years.

Table 6: Future Spending

Given a constant rate of spending, a projected growth rate γ, and an experience slope k, Table 6 gives the factor which should be multiplied by this year's spending to obtain next year's spending.

Next year's spending = this year's spending × [factor]
↑read in Table 6

Use of Table 6

Select appropriate slope (k) and projected growth rate (γ). Read the corresponding factor.

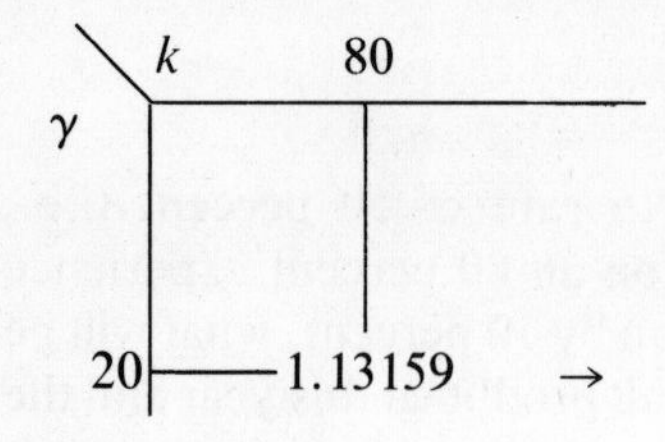

next year's spending will be 13 percent higher than this year

Table 7: Average Cost

Given a projected growth rate γ and an experience slope k, Table 7 gives the factor which should be multiplied by this year's average unit cost to obtain next year's average unit cost.

$$\text{Next year's average cost} = \text{this year's average cost} \times \underset{\text{read in Table 7}}{[\text{factor}]}$$

Use of Table 7

Select appropriate slope (k) and projected growth rate (γ). Read the corresponding factor.

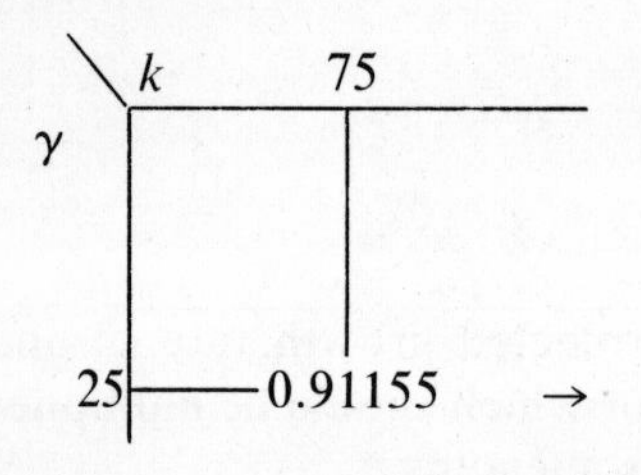

next year's average cost per unit will be 9 percent lower than this year

Appendix: Experience Tables 1 through 7

Table 1
Implicit Price Deflators for U.S. Gross National Product, 1946-1975
[1972 = 1]

Year	*Deflator*	*Year*	*Deflator*
1946	2.25938	1960	1.45624
1947	2.02020	1961	1.44342
1948	1.88218	1962	1.41743
1949	1.90150	1963	1.39684
		1964	1.37533
1950	1.86428		
1951	1.74611	1965	1.34553
1952	1.72414	1966	1.30276
1953	1.69837	1967	1.26550
1954	1.67532	1968	1.21109
		1969	1.15314
1955	1.63988		
1956	1.58983	1970	1.09457
1957	1.53799	1971	1.04145
1958	1.51378	(base) 1972	1.00000
1959	1.48104	1973	0.94410
		1974	0.86058
		1975	0.79145

Table 2
Equivalence between the Slope and the Elasticity Coefficient

k	*Lambda*	*k*	*Lambda*
0.50	1.000000	0.65	0.621488
0.51	0.971431	0.66	0.599462
0.52	0.943416	0.67	0.577767
0.53	0.915936	0.68	0.556393
0.54	0.888969	0.69	0.535331
0.55	0.862496	0.70	0.514573
0.56	0.836501	0.71	0.494109
0.57	0.810966	0.72	0.473931
0.58	0.785875	0.73	0.454032
0.59	0.761213	0.74	0.434403
0.60	0.736965	0.75	0.415037
0.61	0.713119	0.76	0.395929
0.62	0.689660	0.77	0.377069
0.63	0.666576	0.78	0.358454
0.64	0.643856	0.79	0.340075

Table 2 (cont.)

k	*Lambda*	*k*	*Lambda*
0.80	0.321928	0.90	0.152003
0.81	0.304006	0.91	0.136062
0.82	0.286304	0.92	0.120294
0.83	0.268817	0.93	0.104697
0.84	0.251539	0.94	0.089267
0.85	0.234465	0.95	0.074001
0.86	0.217591	0.96	0.058894
0.87	0.200913	0.97	0.043943
0.88	0.184425	0.98	0.029146
0.89	0.168123	0.99	0.014500
		1.00	0.0

Table 3
Growth Needed to Fill Cost Gap

k	λ	$\Delta\rho$ for $dC_n/C_n = 1\%$
1.00	0.0	9999.0000
0.98	0.0291	34.3096
0.96	0.0589	16.9797
0.94	0.0893	11.2023
0.92	0.1203	8.3129
0.90	0.1520	6.5788
0.88	0.1844	5.4223
0.85	0.2176	4.5958
0.84	0.2515	3.9755
0.82	0.2863	3.4928
0.80	0.3219	3.1063
0.78	0.3585	2.7898
0.76	0.3959	2.5257
0.74	0.4344	2.3020
0.72	0.4739	2.1100
0.70	0.5146	1.9434
0.68	0.5564	1.7973
0.66	0.5995	1.6682
0.64	0.6439	1.5531
0.62	0.6897	1.4500
0.60	0.7370	1.3569
0.58	0.7859	1.2725
0.56	0.8365	1.1955
0.54	0.8890	1.1249
0.52	0.9434	1.0600
0.50	1.0000	1.0000

Table 4
Future Costs

P=0.05, *K*=0.70

YEAR* \ * GROWTH	0	5	10	15	20	25	30	35	40	45	50	55	60	65	70
0	1.0000	1.0000	1.0000	1.0000	1.0000	1.0000	1.0000	1.0000	1.0000	1.0000	1.0000	1.0000	1.0000	1.0000	1.0000
1	0.9752	0.9740	0.9728	0.9716	0.9705	0.9693	0.9681	0.9669	0.9658	0.9646	0.9635	0.9623	0.9612	0.9600	0.9589
2	0.9521	0.9488	0.9453	0.9418	0.9382	0.9345	0.9308	0.9270	0.9232	0.9193	0.9154	0.9114	0.9073	0.9033	0.8991
3	0.9306	0.9242	0.9175	0.9106	0.9033	0.8959	0.8881	0.8801	0.8719	0.8635	0.8549	0.8461	0.8371	0.8280	0.8187
4	0.9104	0.9004	0.8896	0.8782	0.8662	0.8535	0.8404	0.8266	0.8124	0.7978	0.7828	0.7675	0.7519	0.7361	0.7201
5	0.8915	0.8772	0.8615	0.8449	0.8270	0.8081	0.7883	0.7676	0.7462	0.7243	0.7020	0.6794	0.6566	0.6339	0.6113
6	0.8737	0.8546	0.8336	0.8108	0.7862	0.7602	0.7329	0.7046	0.6756	0.6462	0.6167	0.5873	0.5583	0.5300	0.5025
7	0.8569	0.8327	0.8057	0.7762	0.7443	0.7107	0.6756	0.6396	0.6033	0.5672	0.5317	0.4972	0.4641	0.4325	0.4025
8	0.8410	0.8113	0.7780	0.7413	0.7018	0.6603	0.6176	0.5746	0.5321	0.4908	0.4513	0.4139	0.3789	0.3464	0.3166
9	0.8260	0.7906	0.7505	0.7064	0.6591	0.6101	0.5604	0.5115	0.4643	0.4197	0.3782	0.3400	0.3053	0.2740	0.2459
10	0.8117	0.7704	0.7233	0.6716	0.6168	0.5607	0.5051	0.4517	0.4016	0.3556	0.3140	0.2768	0.2438	0.2148	0.1894

P=0.05, *K*=0.75

YEAR* \ * GROWTH	0	5	10	15	20	25	30	35	40	45	50	55	60	65	70
0	1.0000	1.0000	1.0000	1.0000	1.0000	1.0000	1.0000	1.0000	1.0000	1.0000	1.0000	1.0000	1.0000	1.0000	1.0000
1	0.9800	0.9790	0.9780	0.9771	0.9761	0.9752	0.9742	0.9733	0.9723	0.9714	0.9704	0.9595	0.9686	0.9676	0.9667
2	0.9613	0.9535	0.9556	0.9528	0.9498	0.9469	0.9438	0.9407	0.9376	0.9344	0.9312	0.9279	0.9246	0.9212	0.9178
3	0.9436	0.9384	0.9329	0.9272	0.9213	0.9151	0.9087	0.9021	0.8953	0.8884	0.8812	0.8739	0.8664	0.8588	0.8510
4	0.9271	0.9188	0.9100	0.9005	0.8906	0.8801	0.8691	0.8577	0.8458	0.8335	0.8208	0.8078	0.7945	0.7810	0.7674
5	0.9115	0.8997	0.8868	0.8729	0.8580	0.8421	0.8254	0.8079	0.7897	0.7709	0.7517	0.7321	0.7123	0.6923	0.6723
6	0.8968	0.8810	0.8635	0.8443	0.8237	0.8016	0.7783	0.7540	0.7289	0.7032	0.6771	0.6510	0.6250	0.5993	0.5740
7	0.8829	0.8627	0.8401	0.8151	0.7881	0.7592	0.7288	0.6974	0.6653	0.6330	0.6008	0.5692	0.5384	0.5086	0.4800
8	0.8697	0.8449	0.8167	0.7855	0.7516	0.7155	0.6780	0.6396	0.6012	0.5633	0.5263	0.4909	0.4571	0.4253	0.3955
9	0.8571	0.8273	0.7933	0.7555	0.7145	0.6713	0.6269	0.5823	0.5386	0.4965	0.4565	0.4189	0.3841	0.3520	0.3225
10	0.8451	0.8102	0.7701	0.7254	0.6772	0.6271	0.5765	0.5268	0.4791	0.4343	0.3928	0.3548	0.3204	0.2893	0.2614

GROWTH

YEAR#	0	5	10	15	20	25	30	35	40	45	50	55	60	65	70
0	1.0000	1.0000	1.0000	1.0000	1.0000	1.0000	1.0000	1.0000	1.0000	1.0000	1.0000	1.0000	1.0000	1.0000	1.0000
1	0.9844	0.9837	0.9829	0.9822	0.9814	0.9807	0.9799	0.9792	0.9785	0.9777	0.9770	0.9763	0.9755	0.9748	0.9741
2	0.9698	0.9676	0.9654	0.9632	0.9609	0.9585	0.9561	0.9537	0.9512	0.9487	0.9462	0.9436	0.9410	0.9383	0.9357
3	0.9560	0.9519	0.9476	0.9431	0.9384	0.9335	0.9285	0.9232	0.9178	0.9123	0.9066	0.9007	0.8947	0.8886	0.8824
4	0.9430	0.9354	0.9294	0.9220	0.9140	0.9057	0.8969	0.8877	0.8781	0.8682	0.8580	0.8474	0.8366	0.8256	0.8143
5	0.9307	0.9213	0.9110	0.8999	0.8880	0.8752	0.8617	0.8475	0.8327	0.8173	0.8014	0.7852	0.7686	0.7519	0.7350
6	0.9190	0.9064	0.8924	0.8770	0.8603	0.8424	0.8233	0.8033	0.7824	0.7610	0.7390	0.7168	0.6945	0.6722	0.6501
7	0.9079	0.8917	0.8736	0.8534	0.8313	0.8076	0.7824	0.7561	0.7290	0.7013	0.6736	0.6459	0.6186	0.5919	0.5659
8	0.8973	0.8774	0.8545	0.8292	0.8013	0.7713	0.7397	0.7071	0.6739	0.6407	0.6079	0.5758	0.5449	0.5152	0.4869
9	0.8873	0.8633	0.8356	0.8045	0.7705	0.7341	0.6961	0.6574	0.6188	0.5809	0.5443	0.5092	0.4761	0.4449	0.4157
10	0.8776	0.8494	0.8165	0.7796	0.7391	0.6963	0.6523	0.6083	0.5651	0.5237	0.4844	0.4477	0.4136	0.3821	0.3532

P=0.05
K=0.80

YEAR#	0	5	10	15	20	25	30	35	40	45	50	55	60	65	70
0	1.0000	1.0000	1.0000	1.0000	1.0000	1.0000	1.0000	1.0000	1.0000	1.0000	1.0000	1.0000	1.0000	1.0000	1.0000
1	0.9886	0.9881	0.9875	0.9870	0.9864	0.9859	0.9853	0.9848	0.9843	0.9837	0.9832	0.9827	0.9821	0.9816	0.9811
2	0.9779	0.9763	0.9747	0.9730	0.9713	0.9696	0.9679	0.9661	0.9642	0.9624	0.9605	0.9586	0.9567	0.9547	0.9527
3	0.9678	0.9647	0.9615	0.9582	0.9547	0.9511	0.9474	0.9435	0.9395	0.9353	0.9310	0.9267	0.9222	0.9176	0.9129
4	0.9582	0.9533	0.9481	0.9425	0.9366	0.9304	0.9238	0.9169	0.9097	0.9022	0.8944	0.8864	0.8782	0.8697	0.8611
5	0.9490	0.9420	0.9344	0.9261	0.9171	0.9075	0.8973	0.8865	0.8751	0.8633	0.8511	0.8385	0.8256	0.8124	0.7991
6	0.9403	0.9309	0.9204	0.9088	0.8962	0.8826	0.8680	0.8526	0.8364	0.8196	0.8023	0.7847	0.7668	0.7488	0.7308
7	0.9321	0.9199	0.9062	0.8910	0.8741	0.8559	0.8364	0.8158	0.7943	0.7723	0.7499	0.7273	0.7048	0.6825	0.6606
8	0.9241	0.9091	0.8919	0.8725	0.8510	0.8277	0.8029	0.7769	0.7502	0.7230	0.6959	0.6690	0.6426	0.6169	0.5921
9	0.9166	0.8984	0.8774	0.8535	0.8270	0.7984	0.7681	0.7368	0.7050	0.6733	0.6421	0.6117	0.5824	0.5544	0.5277
10	0.9093	0.8879	0.8628	0.8341	0.8024	0.7683	0.7326	0.6962	0.6599	0.6243	0.5899	0.5569	0.5257	0.4962	0.4686

P=0.05
K=0.85

*
 *
  * GROWTH
   *

YEAR*	0	5	10	15	20	25	30	35	40	45	50	55	60	65	70
0	1.0000	1.0000	1.0000	1.0000	1.0000	1.0000	1.0000	1.0000	1.0000	1.0000	1.0000	1.0000	1.0000	1.0000	1.0000
1	0.9926	0.9923	0.9919	0.9915	0.9912	0.9908	0.9905	0.9901	0.9898	0.9894	0.9891	0.9887	0.9884	0.9880	0.9877
2	0.9856	0.9846	0.9835	0.9824	0.9813	0.9802	0.9790	0.9779	0.9767	0.9755	0.9742	0.9730	0.9717	0.9704	0.9691
3	0.9790	0.9770	0.9749	0.9727	0.9704	0.9680	0.9656	0.9630	0.9603	0.9576	0.9547	0.9518	0.9488	0.9458	0.9426
4	0.9727	0.9695	0.9660	0.9624	0.9584	0.9543	0.9499	0.9453	0.9405	0.9355	0.9302	0.9248	0.9192	0.9135	0.9076
5	0.9667	0.9620	0.9569	0.9514	0.9454	0.9390	0.9321	0.9248	0.9172	0.9091	0.9008	0.8921	0.8832	0.8740	0.8647
6	0.9609	0.9546	0.9477	0.9399	0.9314	0.9222	0.9123	0.9018	0.8906	0.8790	0.8669	0.8545	0.8418	0.8290	0.8160
7	0.9554	0.9473	0.9382	0.9279	0.9165	0.9040	0.8906	0.8763	0.8613	0.8458	0.8298	0.8135	0.7971	0.7807	0.7643
8	0.9501	0.9401	0.9285	0.9154	0.9007	0.8846	0.8673	0.8490	0.8300	0.8104	0.7905	0.7706	0.7507	0.7312	0.7119
9	0.9451	0.9329	0.9187	0.9024	0.8842	0.8642	0.8428	0.8203	0.7972	0.7738	0.7503	0.7271	0.7044	0.6822	0.6607
10	0.9402	0.9258	0.9088	0.8891	0.8670	0.8429	0.8173	0.7908	0.7638	0.7368	0.7102	0.6842	0.6591	0.6349	0.6117

****
P=0.05
****
K=0.90
****

YEAR	0	5	10	15	20	25	30	35	40	45	50	55	60	65	70
0	1.0000	1.0000	1.0000	1.0000	1.0000	1.0000	1.0000	1.0000	1.0000	1.0000	1.0000	1.0000	1.0000	1.0000	1.0000
1	0.9964	0.9952	0.9962	0.9959	0.9957	0.9955	0.9954	0.9952	0.9950	0.9948	0.9947	0.9945	0.9943	0.9942	0.9940
2	0.9930	0.9925	0.9919	0.9914	0.9909	0.9903	0.9897	0.9892	0.9886	0.9880	0.9874	0.9867	0.9861	0.9855	0.9848
3	0.9897	0.9887	0.9877	0.9866	0.9855	0.9843	0.9831	0.9818	0.9805	0.9791	0.9777	0.9762	0.9748	0.9732	0.9716
4	0.9866	0.9850	0.9833	0.9815	0.9796	0.9775	0.9753	0.9730	0.9706	0.9680	0.9654	0.9627	0.9598	0.9569	0.9539
5	0.9836	0.9813	0.9788	0.9760	0.9731	0.9698	0.9664	0.9627	0.9588	0.9547	0.9504	0.9459	0.9413	0.9365	0.9317
6	0.9808	0.9777	0.9742	0.9703	0.9660	0.9613	0.9563	0.9509	0.9452	0.9391	0.9328	0.9263	0.9196	0.9127	0.9058
7	0.9780	0.9740	0.9694	0.9642	0.9584	0.9521	0.9452	0.9378	0.9299	0.9217	0.9132	0.9044	0.8955	0.8864	0.8773
8	0.9754	0.9704	0.9645	0.9579	0.9504	0.9421	0.9331	0.9234	0.9133	0.9027	0.8919	0.8808	0.8697	0.8586	0.8475
9	0.9729	0.9668	0.9596	0.9512	0.9418	0.9314	0.9201	0.9081	0.8955	0.8826	0.8695	0.8563	0.8432	0.8301	0.8173
10	0.9704	0.9632	0.9545	0.9444	0.9329	0.9202	0.9065	0.8920	0.8771	0.8618	0.8465	0.8313	0.8163	0.8016	0.7872

****
P=0.05
****
K=0.95
****

* GROWTH

P=0.10 *K=0.70*

YEAR*	0	5	10	15	20	25	30	35	40	45	50	55	60	65	70
0	1.0000	1.0000	1.0000	1.0000	1.0000	1.0000	1.0000	1.0000	1.0000	1.0000	1.0000	1.0000	1.0000	1.0000	1.0000
1	0.9521	0.9499	0.9477	0.9455	0.9434	0.9412	0.9390	0.9369	0.9348	0.9327	0.9306	0.9285	0.9265	0.9244	0.9224
2	0.9104	0.9046	0.8986	0.8925	0.8864	0.8803	0.8741	0.8678	0.8615	0.8552	0.8489	0.8425	0.8361	0.8297	0.8233
3	0.8737	0.8632	0.8523	0.8412	0.8299	0.8183	0.8065	0.7946	0.7826	0.7704	0.7582	0.7459	0.7336	0.7214	0.7091
4	0.8410	0.8252	0.8088	0.7918	0.7743	0.7563	0.7381	0.7196	0.7010	0.6823	0.6636	0.6450	0.6265	0.6083	0.5903
5	0.8117	0.7902	0.7677	0.7443	0.7201	0.6954	0.6703	0.6451	0.6199	0.5950	0.5703	0.5462	0.5225	0.4997	0.4775
6	0.7852	0.7578	0.7289	0.6988	0.6678	0.6363	0.6047	0.5732	0.5422	0.5120	0.4828	0.4547	0.4278	0.4023	0.3781
7	0.7611	0.7277	0.6923	0.6554	0.6178	0.5798	0.5422	0.5054	0.4699	0.4359	0.4038	0.3736	0.3454	0.3192	0.2950
8	0.7390	0.6995	0.6576	0.6142	0.5702	0.5264	0.4837	0.4428	0.4042	0.3681	0.3347	0.3041	0.2762	0.2509	0.2280
9	0.7187	0.6732	0.6249	0.5751	0.5252	0.4764	0.4297	0.3860	0.3456	0.3089	0.2757	0.2459	0.2194	0.1959	0.1751
10	0.7000	0.6484	0.5939	0.5381	0.4829	0.4300	0.3805	0.3351	0.2943	0.2580	0.2260	0.1980	0.1736	0.1524	0.1340

P=0.10 *K=0.75*

YEAR*	0	5	10	15	20	25	30	35	40	45	50	55	60	65	70
0	1.0000	1.0000	1.0000	1.0000	1.0000	1.0000	1.0000	1.0000	1.0000	1.0000	1.0000	1.0000	1.0000	1.0000	1.0000
1	0.9612	0.9594	0.9576	0.9558	0.9541	0.9523	0.9505	0.9488	0.9471	0.9454	0.9436	0.9419	0.9403	0.9386	0.9369
2	0.9271	0.9223	0.9174	0.9124	0.9073	0.9023	0.8971	0.8919	0.8867	0.8815	0.8762	0.8709	0.8656	0.8602	0.8549
3	0.8968	0.8881	0.8791	0.8698	0.8603	0.8507	0.8408	0.8307	0.8206	0.8103	0.7999	0.7894	0.7789	0.7584	0.7579
4	0.8697	0.8565	0.8427	0.8283	0.8135	0.7983	0.7827	0.7669	0.7508	0.7347	0.7184	0.7021	0.6858	0.6697	0.6537
5	0.8451	0.8270	0.8080	0.7880	0.7673	0.7460	0.7243	0.7022	0.6800	0.6578	0.6358	0.6140	0.5925	0.5715	0.5509
6	0.8228	0.7995	0.7749	0.7490	0.7221	0.6945	0.6665	0.6383	0.6104	0.5828	0.5558	0.5295	0.5042	0.4798	0.4564
7	0.8023	0.7738	0.7433	0.7113	0.6781	0.6443	0.6103	0.5767	0.5438	0.5119	0.4812	0.4520	0.4243	0.3981	0.3736
8	0.7835	0.7496	0.7132	0.6749	0.6356	0.5960	0.5567	0.5184	0.4816	0.4466	0.4136	0.3829	0.3543	0.3278	0.3034
9	0.7661	0.7267	0.6844	0.6400	0.5948	0.5499	0.5060	0.4640	0.4245	0.3877	0.3537	0.3226	0.2942	0.2685	0.2453
10	0.7500	0.7051	0.6568	0.6066	0.5559	0.5062	0.4587	0.4140	0.3728	0.3353	0.3013	0.2708	0.2436	0.2193	0.1976

GROWTH

P=0.10, K=0.80

YEAR	0	5	10	15	20	25	30	35	40	45	50	55	60	65	70
0	1.0000	1.0000	1.0000	1.0000	1.0000	1.0000	1.0000	1.0000	1.0000	1.0000	1.0000	1.0000	1.0000	1.0000	1.0000
1	0.9698	0.9684	0.9670	0.9656	0.9642	0.9628	0.9614	0.9601	0.9587	0.9573	0.9560	0.9547	0.9533	0.9520	0.9507
2	0.9430	0.9392	0.9353	0.9313	0.9274	0.9233	0.9192	0.9151	0.9110	0.9068	0.9026	0.8983	0.8941	0.8898	0.8855
3	0.9190	0.9121	0.9049	0.8975	0.8899	0.8821	0.8741	0.8660	0.8578	0.8494	0.8410	0.8324	0.8238	0.8152	0.8065
4	0.8973	0.8868	0.8757	0.8641	0.8521	0.8397	0.8270	0.8139	0.8007	0.7873	0.7737	0.7551	0.7464	0.7327	0.7191
5	0.8776	0.8630	0.8476	0.8313	0.8143	0.7967	0.7786	0.7602	0.7415	0.7226	0.7038	0.6850	0.6663	0.6479	0.6298
6	0.8596	0.8407	0.8205	0.7991	0.7768	0.7537	0.7300	0.7060	0.6819	0.6578	0.6341	0.6107	0.5879	0.5657	0.5442
7	0.8430	0.8196	0.7945	0.7678	0.7398	0.7111	0.6818	0.6525	0.6234	0.5949	0.5670	0.5401	0.5143	0.4895	0.4659
8	0.8276	0.7997	0.7694	0.7372	0.7036	0.6693	0.6349	0.6007	0.5674	0.5351	0.5042	0.4749	0.4471	0.4210	0.3965
9	0.8133	0.7807	0.7451	0.7074	0.6684	0.6288	0.5896	0.5513	0.5145	0.4795	0.4466	0.4158	0.3872	0.3507	0.3362
10	0.8000	0.7625	0.7218	0.6786	0.6342	0.5898	0.5463	0.5046	0.4652	0.4284	0.3944	0.3630	0.3344	0.3082	0.2843

P=0.10, K=0.85

YEAR	0	5	10	15	20	25	30	35	40	45	50	55	60	65	70
0	1.0000	1.0000	1.0000	1.0000	1.0000	1.0000	1.0000	1.0000	1.0000	1.0000	1.0000	1.0000	1.0000	1.0000	1.0000
1	0.9779	0.9769	0.9758	0.9748	0.9738	0.9728	0.9718	0.9707	0.9697	0.9688	0.9678	0.9568	0.9658	0.9648	0.9639
2	0.9582	0.9553	0.9524	0.9495	0.9466	0.9435	0.9405	0.9374	0.9343	0.9312	0.9281	0.9249	0.9217	0.9185	0.9152
3	0.9403	0.9352	0.9298	0.9242	0.9185	0.9127	0.9067	0.9005	0.8943	0.8879	0.8815	0.8750	0.8684	0.8617	0.8550
4	0.9241	0.9162	0.9078	0.8991	0.8900	0.8805	0.8708	0.8608	0.8505	0.8401	0.8296	0.8189	0.8081	0.7973	0.7865
5	0.9093	0.8983	0.8865	0.8741	0.8610	0.8474	0.8334	0.8190	0.8042	0.7893	0.7742	0.7591	0.7440	0.7290	0.7141
6	0.8957	0.8813	0.8658	0.8493	0.8320	0.8138	0.7951	0.7760	0.7566	0.7371	0.7176	0.6983	0.6792	0.6604	0.6420
7	0.8830	0.8651	0.8457	0.8249	0.8029	0.7801	0.7566	0.7328	0.7088	0.6850	0.6615	0.6385	0.6161	0.5944	0.5734
8	0.8713	0.8497	0.8262	0.8008	0.7741	0.7465	0.7183	0.6899	0.6618	0.6342	0.6073	0.5814	0.5564	0.5326	0.5098
9	0.8603	0.8350	0.8071	0.7772	0.7457	0.7133	0.6806	0.6481	0.6163	0.5855	0.5559	0.5277	0.5010	0.4758	0.4520
10	0.8500	0.8209	0.7886	0.7540	0.7177	0.6807	0.6438	0.6077	0.5727	0.5394	0.5078	0.4781	0.4503	0.4243	0.4002

*
 *
  * GROWTH
   *

P=0.10 *K=0.90*

YEAR*	0	5	10	15	20	25	30	35	40	45	50	55	60	65	70
0	1.0000	1.0000	1.0000	1.0000	1.0000	1.0000	1.0000	1.0000	1.0000	1.0000	1.0000	1.0000	1.0000	1.0000	1.0000
1	0.9856	0.9849	0.9843	0.9836	0.9829	0.9823	0.9816	0.9809	0.9803	0.9796	0.9790	0.9783	0.9777	0.9771	0.9764
2	0.9727	0.9708	0.9689	0.9670	0.9650	0.9630	0.9610	0.9590	0.9569	0.9548	0.9527	0.9506	0.9485	0.9464	0.9442
3	0.9609	0.9575	0.9539	0.9502	0.9464	0.9425	0.9385	0.9343	0.9301	0.9258	0.9215	0.9171	0.9126	0.9080	0.9034
4	0.9501	0.9448	0.9392	0.9334	0.9272	0.9208	0.9142	0.9074	0.9004	0.8932	0.8859	0.8785	0.8710	0.8634	0.8558
5	0.9402	0.9328	0.9249	0.9164	0.9076	0.8982	0.8886	0.8786	0.8683	0.8578	0.8471	0.8364	0.8256	0.8147	0.8039
6	0.9311	0.9213	0.9108	0.8995	0.8876	0.8750	0.8619	0.8484	0.8346	0.8206	0.8065	0.7923	0.7782	0.7642	0.7503
7	0.9225	0.9104	0.8971	0.8827	0.8674	0.8513	0.8346	0.8174	0.8000	0.7825	0.7650	0.7476	0.7305	0.7137	0.6973
8	0.9145	0.8998	0.8836	0.8659	0.8471	0.8273	0.8069	0.7861	0.7652	0.7444	0.7238	0.7035	0.6838	0.6647	0.6461
9	0.9070	0.8897	0.8703	0.8492	0.8268	0.8033	0.7792	0.7549	0.7307	0.7068	0.6834	0.6508	0.6389	0.6178	0.5977
10	0.9000	0.8799	0.8573	0.8327	0.8065	0.7793	0.7517	0.7240	0.6967	0.6702	0.6445	0.6198	0.5961	0.5736	0.5522

P=0.10 *K=0.95*

YEAR*	0	5	10	15	20	25	30	35	40	45	50	55	60	65	70
0	1.0000	1.0000	1.0000	1.0000	1.0000	1.0000	1.0000	1.0000	1.0000	1.0000	1.0000	1.0000	1.0000	1.0000	1.0000
1	0.9930	0.9926	0.9923	0.9920	0.9916	0.9913	0.9910	0.9907	0.9904	0.9900	0.9897	0.9894	0.9891	0.9888	0.9884
2	0.9866	0.9857	0.9847	0.9838	0.9828	0.9818	0.9808	0.9798	0.9788	0.9778	0.9767	0.9757	0.9746	0.9735	0.9724
3	0.9808	0.9791	0.9773	0.9754	0.9735	0.9716	0.9696	0.9675	0.9654	0.9632	0.9610	0.9587	0.9564	0.9541	0.9518
4	0.9754	0.9728	0.9699	0.9670	0.9639	0.9606	0.9573	0.9538	0.9502	0.9465	0.9427	0.9389	0.9350	0.9310	0.9270
5	0.9704	0.9667	0.9627	0.9584	0.9539	0.9491	0.9441	0.9389	0.9336	0.9280	0.9224	0.9167	0.9109	0.9050	0.8992
6	0.9658	0.9609	0.9555	0.9498	0.9436	0.9371	0.9302	0.9231	0.9157	0.9082	0.9006	0.8928	0.8851	0.8773	0.8695
7	0.9615	0.9553	0.9485	0.9411	0.9331	0.9246	0.9157	0.9065	0.8971	0.8875	0.8777	0.8680	0.8582	0.8486	0.8390
8	0.9574	0.9499	0.9415	0.9323	0.9224	0.9118	0.9008	0.8895	0.8779	0.8661	0.8544	0.8427	0.8311	0.8197	0.8084
9	0.9536	0.9447	0.9346	0.9235	0.9115	0.8988	0.8856	0.8721	0.8583	0.8445	0.8308	0.8173	0.8040	0.7910	0.7783
10	0.9500	0.9396	0.9278	0.9147	0.9006	0.8857	0.8703	0.8545	0.8387	0.8230	0.8074	0.7922	0.7774	0.7630	0.7490

* GROWTH

YEAR*	0	5	10	15	20	25	30	35	40	45	50	55	60	65	70
0	1.0000	1.0000	1.0000	1.0000	1.0000	1.0000	1.0000	1.0000	1.0000	1.0000	1.0000	1.0000	1.0000	1.0000	1.0000
1	0.9306	0.9275	0.9244	0.9214	0.9184	0.9154	0.9124	0.9095	0.9066	0.9037	0.9008	0.8980	0.8952	0.8924	0.8897
2	0.8737	0.8659	0.8581	0.8502	0.8423	0.8343	0.8264	0.8185	0.8106	0.8027	0.7948	0.7870	0.7792	0.7714	0.7637
3	0.8260	0.8127	0.7991	0.7854	0.7716	0.7577	0.7437	0.7297	0.7157	0.7018	0.6879	0.6742	0.6606	0.6471	0.6339
4	0.7852	0.7660	0.7463	0.7264	0.7062	0.6858	0.6655	0.6452	0.6251	0.6053	0.5857	0.5565	0.5477	0.5294	0.5116
5	0.7498	0.7246	0.6987	0.6723	0.6457	0.6191	0.5927	0.5667	0.5411	0.5162	0.4921	0.4588	0.4464	0.4249	0.4044
6	0.7187	0.6875	0.6554	0.6228	0.5901	0.5577	0.5258	0.4949	0.4650	0.4365	0.4093	0.3835	0.3593	0.3365	0.3152
7	0.6912	0.6540	0.6158	0.5773	0.5389	0.5013	0.4650	0.4303	0.3974	0.3666	0.3379	0.3112	0.2867	0.2541	0.2434
8	0.6665	0.6236	0.5795	0.5353	0.4919	0.4500	0.4101	0.3728	0.3382	0.3064	0.2774	0.2511	0.2274	0.2060	0.1868
9	0.6443	0.5957	0.5461	0.4967	0.4488	0.4034	0.3610	0.3221	0.2869	0.2552	0.2270	0.2019	0.1797	0.1602	0.1429
10	0.6241	0.5700	0.5151	0.4611	0.4094	0.3612	0.3172	0.2778	0.2428	0.2120	0.1852	0.1519	0.1417	0.1242	0.1091

P=0.15
K=0.70

YEAR*	0	5	10	15	20	25	30	35	40	45	50	55	60	65	70
0	1.0000	1.0000	1.0000	1.0000	1.0000	1.0000	1.0000	1.0000	1.0000	1.0000	1.0000	1.0000	1.0000	1.0000	1.0000
1	0.9436	0.9411	0.9386	0.9361	0.9336	0.9312	0.9287	0.9263	0.9239	0.9216	0.9192	0.9169	0.9146	0.9123	0.9100
2	0.8968	0.8904	0.8838	0.8773	0.8707	0.8641	0.8575	0.8508	0.8442	0.8375	0.8309	0.8243	0.8177	0.8111	0.8046
3	0.8571	0.8459	0.8346	0.8230	0.8113	0.7994	0.7875	0.7755	0.7635	0.7515	0.7396	0.7276	0.7158	0.7040	0.6923
4	0.8228	0.8065	0.7898	0.7727	0.7553	0.7377	0.7200	0.7023	0.6846	0.6670	0.6495	0.6323	0.6154	0.5987	0.5824
5	0.7927	0.7712	0.7489	0.7260	0.7028	0.6793	0.6558	0.6325	0.6094	0.5867	0.5644	0.5428	0.5218	0.5014	0.4818
6	0.7661	0.7392	0.7112	0.6825	0.6535	0.6243	0.5954	0.5670	0.5393	0.5124	0.4865	0.4616	0.4379	0.4154	0.3941
7	0.7424	0.7100	0.6764	0.6420	0.6074	0.5730	0.5392	0.5065	0.4751	0.4451	0.4168	0.3901	0.3650	0.3417	0.3199
8	0.7209	0.6832	0.6440	0.6041	0.5643	0.5251	0.4873	0.4512	0.4171	0.3852	0.3555	0.3281	0.3029	0.2797	0.2585
9	0.7014	0.6585	0.6139	0.5687	0.5241	0.4808	0.4396	0.4010	0.3652	0.3324	0.3024	0.2751	0.2505	0.2283	0.2082
10	0.6837	0.6355	0.5855	0.5355	0.4866	0.4398	0.3961	0.3559	0.3192	0.2862	0.2566	0.2303	0.2068	0.1860	0.1675

P=0.15
K=0.75

GROWTH

YEAR	0	5	10	15	20	25	30	35	40	45	50	55	60	65	70
0	1.0000	1.0000	1.0000	1.0000	1.0000	1.0000	1.0000	1.0000	1.0000	1.0000	1.0000	1.0000	1.0000	1.0000	1.0000
1	0.9560	0.9540	0.9520	0.9501	0.9481	0.9462	0.9443	0.9424	0.9405	0.9386	0.9368	0.9349	0.9331	0.9313	0.9295
2	0.9190	0.9139	0.9087	0.9034	0.8982	0.8929	0.8876	0.8822	0.8769	0.8715	0.8662	0.8508	0.8555	0.8501	0.8448
3	0.8873	0.8783	0.8691	0.8598	0.8502	0.8406	0.8309	0.8211	0.8112	0.8013	0.7913	0.7314	0.7715	0.7617	0.7518
4	0.8596	0.8464	0.8327	0.8137	0.8044	0.7898	0.7751	0.7602	0.7453	0.7304	0.7156	0.7008	0.6862	0.6717	0.6575
5	0.8351	0.8175	0.7991	0.7801	0.7606	0.7409	0.7209	0.7009	0.6810	0.6612	0.6417	0.6225	0.6037	0.5854	0.5676
6	0.8133	0.7910	0.7677	0.7436	0.7189	0.6939	0.6689	0.6440	0.6194	0.5953	0.5718	0.5491	0.5271	0.5059	0.4856
7	0.7937	0.7667	0.7384	0.7091	0.6792	0.6492	0.6194	0.5900	0.5614	0.5337	0.5072	0.4818	0.4576	0.4347	0.4131
8	0.7758	0.7442	0.7108	0.6764	0.6416	0.6068	0.5726	0.5394	0.5075	0.4771	0.4484	0.4213	0.3959	0.3722	0.3501
9	0.7595	0.7232	0.6849	0.6455	0.6058	0.5667	0.5287	0.4923	0.4578	0.4255	0.3954	0.3675	0.3417	0.3180	0.2961
10	0.7445	0.7035	0.6603	0.6161	0.5719	0.5288	0.4876	0.4487	0.4124	0.3789	0.3482	0.3201	0.2945	0.2712	0.2501

P=0.15
K=0.80

YEAR	0	5	10	15	20	25	30	35	40	45	50	55	60	65	70
0	1.0000	1.0000	1.0000	1.0000	1.0000	1.0000	1.0000	1.0000	1.0000	1.0000	1.0000	1.0000	1.0000	1.0000	1.0000
1	0.9678	0.9663	0.9648	0.9634	0.9619	0.9605	0.9591	0.9577	0.9563	0.9549	0.9535	0.9522	0.9508	0.9495	0.9481
2	0.9403	0.9365	0.9326	0.9287	0.9248	0.9208	0.9168	0.9128	0.9087	0.9047	0.9005	0.8966	0.8925	0.8885	0.8844
3	0.9166	0.9098	0.9029	0.8958	0.8886	0.8812	0.8738	0.8662	0.8586	0.8510	0.8433	0.8356	0.8279	0.8201	0.8124
4	0.3957	0.8856	0.8752	0.8644	0.8534	0.8421	0.8306	0.8190	0.8073	0.7955	0.7837	0.7719	0.7601	0.7484	0.7368
5	0.3770	0.8635	0.8493	0.8345	0.8193	0.8038	0.7879	0.7720	0.7559	0.7399	0.7239	0.7081	0.6925	0.6771	0.6620
6	0.8603	0.8430	0.8249	0.8059	0.7864	0.7664	0.7461	0.7258	0.7055	0.6854	0.6656	0.6462	0.6272	0.6088	0.5909
7	0.8451	0.8241	0.8018	0.7785	0.7545	0.7301	0.7055	0.6809	0.6567	0.6330	0.6099	0.5875	0.5659	0.5452	0.5252
8	0.8312	0.8064	0.7799	0.7522	0.7238	0.6950	0.6662	0.6379	0.6102	0.5833	0.5575	0.5328	0.5093	0.4869	0.4656
9	0.8185	0.7897	0.7591	0.7270	0.6942	0.6612	0.6286	0.5968	0.5661	0.5367	0.5088	0.4824	0.4575	0.4341	0.4121
10	0.8067	0.7741	0.7391	0.7027	0.6657	0.6288	0.5927	0.5578	0.5246	0.4932	0.4638	0.4362	0.4105	0.3866	0.3644

P=0.15
K=0.85

GROWTH

P=0.15, *K=0.90*

YEAR	0	5	10	15	20	25	30	35	40	45	50	55	60	65	70
0	1.0000	1.0000	1.0000	1.0000	1.0000	1.0000	1.0000	1.0000	1.0000	1.0000	1.0000	1.0000	1.0000	1.0000	1.0000
1	0.9790	0.9780	0.9771	0.9761	0.9752	0.9742	0.9733	0.9724	0.9714	0.9705	0.9696	0.9687	0.9678	0.9669	0.9661
2	0.9609	0.9584	0.9558	0.9532	0.9506	0.9479	0.9452	0.9425	0.9398	0.9371	0.9344	0.9317	0.9289	0.9262	0.9235
3	0.9451	0.9406	0.9359	0.9311	0.9263	0.9213	0.9162	0.9111	0.9059	0.9007	0.8954	0.8901	0.8847	0.8794	0.8740
4	0.9311	0.9243	0.9172	0.9099	0.9023	0.8946	0.8867	0.8786	0.8704	0.8622	0.8538	0.8455	0.8371	0.8287	0.8204
5	0.9185	0.9092	0.8995	0.8893	0.8788	0.8680	0.8568	0.8455	0.8341	0.8226	0.8110	0.7995	0.7880	0.7766	0.7654
6	0.9070	0.8952	0.8827	0.8695	0.8557	0.8415	0.8271	0.8124	0.7976	0.7828	0.7680	0.7534	0.7391	0.7249	0.7110
7	0.8966	0.8821	0.8665	0.8502	0.8331	0.8155	0.7976	0.7795	0.7614	0.7435	0.7257	0.7084	0.6914	0.6748	0.6587
8	0.8871	0.8698	0.8512	0.8315	0.8109	0.7899	0.7685	0.7472	0.7260	0.7051	0.6847	0.6649	0.6457	0.6271	0.6093
9	0.8782	0.8581	0.8363	0.8133	0.7893	0.7648	0.7401	0.7156	0.6915	0.6680	0.6453	0.6234	0.6023	0.5822	0.5629
10	0.8700	0.8470	0.8221	0.7956	0.7681	0.7402	0.7124	0.6850	0.6582	0.6324	0.6077	0.5840	0.5615	0.5401	0.5198

P=0.15, *K=0.95*

YEAR	0	5	10	15	20	25	30	35	40	45	50	55	60	65	70
0	1.0000	1.0000	1.0000	1.0000	1.0000	1.0000	1.0000	1.0000	1.0000	1.0000	1.0000	1.0000	1.0000	1.0000	1.0000
1	0.9897	0.9892	0.9888	0.9883	0.9878	0.9874	0.9869	0.9864	0.9860	0.9855	0.9851	0.9846	0.9842	0.9838	0.9833
2	0.9808	0.9795	0.9782	0.9769	0.9756	0.9743	0.9730	0.9716	0.9702	0.9689	0.9675	0.9561	0.9648	0.9634	0.9620
3	0.9729	0.9706	0.9683	0.9659	0.9634	0.9609	0.9583	0.9557	0.9530	0.9503	0.9476	0.9449	0.9421	0.9393	0.9365
4	0.9658	0.9624	0.9588	0.9551	0.9512	0.9472	0.9431	0.9389	0.9347	0.9303	0.9260	0.9215	0.9171	0.9126	0.9081
5	0.9594	0.9547	0.9497	0.9445	0.9390	0.9334	0.9275	0.9216	0.9155	0.9093	0.9031	0.8968	0.8905	0.8842	0.8779
6	0.9536	0.9475	0.9410	0.9342	0.9269	0.9194	0.9117	0.9038	0.8957	0.8876	0.8794	0.8713	0.8631	0.8550	0.8470
7	0.9483	0.9408	0.9327	0.9240	0.9149	0.9055	0.8957	0.8858	0.8757	0.8656	0.8555	0.8455	0.8355	0.8257	0.8161
8	0.9433	0.9343	0.9245	0.9141	0.9030	0.8915	0.8797	0.8677	0.8556	0.8436	0.8316	0.8198	0.8082	0.7968	0.7857
9	0.9387	0.9282	0.9167	0.9043	0.8912	0.8776	0.8637	0.8497	0.8356	0.8217	0.8079	0.7945	0.7813	0.7684	0.7560
10	0.9344	0.9224	0.9090	0.8946	0.8795	0.8638	0.8478	0.8318	0.8158	0.8001	0.7847	0.7596	0.7550	0.7409	0.7272

*
 *
 * GROWTH
 *

YEAR*	0	5	10	15	20	25	30	35	40	45	50	55	60	65	70
0	1.0000	1.0000	1.0000	1.0000	1.0000	1.0000	1.0000	1.0000	1.0000	1.0000	1.0000	1.0000	1.0000	1.0000	1.0000
1	0.9104	0.9066	0.9027	0.8990	0.8952	0.8915	0.8879	0.8843	0.8807	0.8772	0.8737	0.8703	0.8669	0.8635	0.8602
2	0.8410	0.8317	0.8225	0.8132	0.8040	0.7948	0.7857	0.7766	0.7676	0.7586	0.7498	0.7410	0.7323	0.7237	0.7152
3	0.7852	0.7700	0.7546	0.7393	0.7239	0.7086	0.6934	0.6782	0.6633	0.6485	0.6339	0.6196	0.6055	0.5917	0.5781
4	0.7390	0.7177	0.6962	0.6747	0.6531	0.6317	0.6106	0.5897	0.5692	0.5492	0.5297	0.5108	0.4924	0.4746	0.4575
5	0.7000	0.6728	0.6452	0.6176	0.5902	0.5632	0.5367	0.5109	0.4860	0.4619	0.4388	0.4168	0.3957	0.3757	0.3568
6	0.6665	0.6335	0.6001	0.5669	0.5341	0.5021	0.4712	0.4415	0.4132	0.3864	0.3612	0.3376	0.3154	0.2948	0.2756
7	0.6373	0.5987	0.5599	0.5214	0.4838	0.4476	0.4132	0.3807	0.3503	0.3221	0.2960	0.2720	0.2501	0.2300	0.2116
8	0.6116	0.5677	0.5236	0.4804	0.4387	0.3991	0.3620	0.3277	0.2963	0.2677	0.2418	0.2185	0.1976	0.1788	0.1619
9	0.5887	0.5397	0.4908	0.4432	0.3980	0.3558	0.3170	0.2818	0.2503	0.2221	0.1972	0.1751	0.1557	0.1386	0.1236
10	0.5682	0.5143	0.4609	0.4095	0.3613	0.3172	0.2774	0.2421	0.2111	0.1840	0.1605	0.1401	0.1226	0.1074	0.0943

P=0.20
K=0.70

YEAR*	0	5	10	15	20	25	30	35	40	45	50	55	60	65	70
0	1.0000	1.0000	1.0000	1.0000	1.0000	1.0000	1.0000	1.0000	1.0000	1.0000	1.0000	1.0000	1.0000	1.0000	1.0000
1	0.9271	0.9239	0.9208	0.9177	0.9146	0.9115	0.9085	0.9056	0.9026	0.8997	0.8968	0.8940	0.8912	0.8884	0.8856
2	0.8697	0.8619	0.8542	0.8464	0.8387	0.8309	0.8232	0.8155	0.8079	0.8003	0.7927	0.7352	0.7778	0.7704	0.7631
3	0.8228	0.8099	0.7969	0.7838	0.7706	0.7574	0.7443	0.7311	0.7181	0.7052	0.6924	0.6797	0.6672	0.6549	0.6428
4	0.7835	0.7653	0.7463	0.7280	0.7092	0.6904	0.6717	0.6531	0.6348	0.6167	0.5990	0.5817	0.5647	0.5482	0.5322
5	0.7500	0.7264	0.7023	0.6780	0.6536	0.6294	0.6054	0.5818	0.5588	0.5363	0.5146	0.4936	0.4734	0.4541	0.4355
6	0.7209	0.6920	0.6624	0.6327	0.6030	0.5737	0.5450	0.5171	0.4903	0.4645	0.4399	0.4165	0.3943	0.3734	0.3536
7	0.6953	0.6612	0.6263	0.5914	0.5568	0.5229	0.4902	0.4589	0.4291	0.4010	0.3746	0.3499	0.3270	0.3056	0.2858
8	0.6726	0.6334	0.5934	0.5536	0.5145	0.4767	0.4407	0.4067	0.3749	0.3455	0.3183	0.2933	0.2704	0.2494	0.2303
9	0.6522	0.6081	0.5632	0.5188	0.4756	0.4345	0.3959	0.3601	0.3272	0.2971	0.2699	0.2453	0.2231	0.2032	0.1852
10	0.6338	0.5849	0.5354	0.4867	0.4400	0.3961	0.3555	0.3186	0.2852	0.2553	0.2286	0.2050	0.1840	0.1653	0.1489

P=0.20
K=0.75

GROWTH

ρ = 0.20, K = 0.80

YEAR*	0	5	10	15	20	25	30	35	40	45	50	55	60	65	70
0	1.0000	1.0000	1.0000	1.0000	1.0000	1.0000	1.0000	1.0000	1.0000	1.0000	1.0000	1.0000	1.0000	1.0000	1.0000
1	0.9430	0.9405	0.9380	0.9355	0.9331	0.9307	0.9283	0.9259	0.9236	0.9213	0.9190	0.9167	0.9145	0.9123	0.9101
2	0.8973	0.8911	0.8849	0.8787	0.8724	0.8662	0.8599	0.8537	0.8475	0.8413	0.8351	0.8290	0.8229	0.8169	0.8109
3	0.8596	0.8491	0.8385	0.8278	0.8170	0.8061	0.7952	0.7844	0.7735	0.7627	0.7519	0.7412	0.7306	0.7201	0.7098
4	0.8276	0.8126	0.7973	0.7818	0.7661	0.7503	0.7344	0.7186	0.7029	0.6874	0.6720	0.6569	0.6420	0.6274	0.6131
5	0.8000	0.7804	0.7602	0.7397	0.7190	0.6982	0.6775	0.6570	0.6367	0.6168	0.5973	0.5784	0.5599	0.5420	0.5248
6	0.7758	0.7515	0.7265	0.7011	0.6755	0.6498	0.6245	0.5996	0.5753	0.5517	0.5289	0.5069	0.4859	0.4657	0.4465
7	0.7544	0.7255	0.6957	0.6653	0.6349	0.6048	0.5752	0.5465	0.5188	0.4922	0.4669	0.4429	0.4202	0.3987	0.3785
8	0.7352	0.7017	0.6671	0.6321	0.5972	0.5629	0.5296	0.4976	0.4672	0.4385	0.4115	0.3862	0.3626	0.3406	0.3202
9	0.7179	0.6799	0.6406	0.6011	0.5619	0.5239	0.4874	0.4528	0.4204	0.3901	0.3621	0.3362	0.3124	0.2905	0.2704
10	0.7021	0.6597	0.6159	0.5720	0.5290	0.4876	0.4484	0.4118	0.3779	0.3468	0.3184	0.2925	0.2689	0.2476	0.2282

ρ = 0.20, K = 0.85

YEAR*	0	5	10	15	20	25	30	35	40	45	50	55	60	65	70
0	1.0000	1.0000	1.0000	1.0000	1.0000	1.0000	1.0000	1.0000	1.0000	1.0000	1.0000	1.0000	1.0000	1.0000	1.0000
1	0.9582	0.9563	0.9544	0.9526	0.9508	0.9490	0.9473	0.9455	0.9438	0.9420	0.9403	0.9387	0.9370	0.9353	0.9337
2	0.9241	0.9195	0.9148	0.9101	0.9054	0.9006	0.8959	0.8912	0.8865	0.8817	0.8770	0.8723	0.8677	0.8630	0.8584
3	0.8957	0.8877	0.8796	0.8714	0.8631	0.8547	0.8463	0.8379	0.8294	0.8209	0.8125	0.8040	0.7956	0.7873	0.7790
4	0.8713	0.8597	0.8479	0.8358	0.8236	0.8112	0.7987	0.7861	0.7736	0.7611	0.7486	0.7363	0.7241	0.7121	0.7002
5	0.8500	0.8348	0.8190	0.8029	0.7864	0.7698	0.7531	0.7364	0.7198	0.7033	0.6871	0.6711	0.6555	0.6402	0.6252
6	0.8312	0.8122	0.7924	0.7721	0.7514	0.7306	0.7097	0.6890	0.6685	0.6484	0.6288	0.6097	0.5911	0.5732	0.5559
7	0.8144	0.7916	0.7677	0.7432	0.7183	0.6933	0.6685	0.6440	0.6201	0.5968	0.5743	0.5526	0.5318	0.5119	0.4928
8	0.7993	0.7726	0.7447	0.7160	0.6870	0.6580	0.6294	0.6015	0.5745	0.5486	0.5237	0.5001	0.4776	0.4564	0.4363
9	0.7855	0.7550	0.7230	0.6902	0.6572	0.6245	0.5925	0.5616	0.5320	0.5038	0.4772	0.4521	0.4285	0.4065	0.3858
10	0.7729	0.7386	0.7026	0.6658	0.6289	0.5926	0.5576	0.5240	0.4923	0.4624	0.4345	0.4085	0.3842	0.3618	0.3409

*
*
* GROWTH
*

YEAR*	0	5	10	15	20	25	30	35	40	45	50	55	60	65	70
0	1.0000	1.0000	1.0000	1.0000	1.0000	1.0000	1.0000	1.0000	1.0000	1.0000	1.0000	1.0000	1.0000	1.0000	1.0000
1	0.9727	0.9714	0.9702	0.9690	0.9678	0.9667	0.9655	0.9643	0.9632	0.9620	0.9609	0.9598	0.9587	0.9576	0.9565
2	0.9501	0.9470	0.9439	0.9408	0.9376	0.9344	0.9312	0.9280	0.9248	0.9216	0.9185	0.9153	0.9121	0.9089	0.9057
3	0.9311	0.9257	0.9202	0.9146	0.9090	0.9033	0.8975	0.8916	0.8858	0.8799	0.8740	0.8681	0.8623	0.8564	0.8506
4	0.9145	0.9067	0.8986	0.8903	0.8818	0.8731	0.8644	0.8556	0.8467	0.8378	0.8289	0.8200	0.8112	0.8024	0.7937
5	0.9000	0.8895	0.8786	0.8673	0.8558	0.8440	0.8321	0.8201	0.8080	0.7960	0.7840	0.7722	0.7605	0.7489	0.7375
6	0.8871	0.8738	0.8600	0.8456	0.8309	0.8159	0.8007	0.7854	0.7702	0.7551	0.7402	0.7256	0.7112	0.6971	0.6834
7	0.8754	0.8594	0.8425	0.8250	0.8070	0.7887	0.7702	0.7518	0.7336	0.7156	0.6980	0.6808	0.6640	0.6478	0.6321
8	0.3648	0.8460	0.8260	0.8053	0.7839	0.7624	0.7407	0.7193	0.6982	0.6775	0.6575	0.6381	0.6194	0.6014	0.5840
9	0.8551	0.8334	0.8104	0.7864	0.7617	0.7369	0.7122	0.6879	0.6642	0.6412	0.6190	0.5977	0.5773	0.5579	0.5393
10	0.8462	0.8216	0.7955	0.7682	0.7403	0.7124	0.6847	0.6577	0.6316	0.6065	0.5825	0.5596	0.5379	0.5173	0.4978

P=0.20
K=0.90

YEAR	0	5	10	15	20	25	30	35	40	45	50	55	60	65	70
0	1.0000	1.0000	1.0000	1.0000	1.0000	1.0000	1.0000	1.0000	1.0000	1.0000	1.0000	1.0000	1.0000	1.0000	1.0000
1	0.9866	0.9860	0.9854	0.9848	0.9842	0.9836	0.9830	0.9825	0.9819	0.9813	0.9808	0.9802	0.9797	0.9791	0.9786
2	0.9754	0.9739	0.9723	0.9707	0.9691	0.9675	0.9659	0.9643	0.9627	0.9611	0.9594	0.9578	0.9562	0.9546	0.9529
3	0.9658	0.9631	0.9603	0.9575	0.9546	0.9517	0.9487	0.9457	0.9427	0.9396	0.9366	0.9335	0.9304	0.9273	0.9242
4	0.9574	0.9534	0.9493	0.9450	0.9406	0.9361	0.9315	0.9269	0.9222	0.9174	0.9127	0.9079	0.9031	0.8984	0.8936
5	0.9500	0.9446	0.9389	0.9331	0.9270	0.9208	0.9144	0.9079	0.9014	0.8949	0.8883	0.8817	0.8752	0.8687	0.8622
6	0.9433	0.9365	0.9292	0.9216	0.9138	0.9057	0.8974	0.8891	0.8806	0.8722	0.8638	0.8554	0.8471	0.8389	0.8308
7	0.9373	0.9289	0.9200	0.9106	0.9009	0.8908	0.8806	0.8703	0.8600	0.8497	0.8394	0.8293	0.8193	0.8095	0.7999
8	0.9317	0.9218	0.9112	0.8999	0.8883	0.8763	0.8641	0.8518	0.8395	0.8274	0.8154	0.8035	0.7920	0.7807	0.7697
9	0.9266	0.9151	0.9027	0.8896	0.8759	0.8619	0.8477	0.8335	0.8194	0.8054	0.7917	0.7784	0.7653	0.7527	0.7404
10	0.9219	0.9088	0.8946	0.8795	0.8638	0.8478	0.8316	0.8155	0.7996	0.7839	0.7687	0.7538	0.7394	0.7255	0.7120

P=0.20
K=0.95

GROWTH

P=0.25, K=0.70

YEAR	0	5	10	15	20	25	30	35	40	45	50	55	60	65	70
0	1.0000	1.0000	1.0000	1.0000	1.0000	1.0000	1.0000	1.0000	1.0000	1.0000	1.0000	1.0000	1.0000	1.0000	1.0000
1	0.8915	0.8870	0.8825	0.8781	0.8737	0.8694	0.8652	0.8610	0.8569	0.8529	0.8489	0.8449	0.8410	0.8372	0.8334
2	0.8117	0.8013	0.7909	0.7806	0.7704	0.7603	0.7503	0.7405	0.7307	0.7210	0.7115	0.7021	0.6929	0.6838	0.6748
3	0.7498	0.7332	0.7167	0.7003	0.6840	0.6678	0.6519	0.6362	0.6208	0.6056	0.5908	0.5762	0.5620	0.5482	0.5347
4	0.7000	0.6775	0.6549	0.6325	0.6103	0.5885	0.5671	0.5462	0.5259	0.5062	0.4871	0.4686	0.4509	0.4338	0.4174
5	0.6588	0.6305	0.6023	0.5743	0.5468	0.5199	0.4939	0.4688	0.4446	0.4216	0.3996	0.3787	0.3589	0.3402	0.3226
6	0.6241	0.5903	0.5567	0.5236	0.4913	0.4602	0.4304	0.4021	0.3754	0.3502	0.3267	0.3048	0.2843	0.2654	0.2478
7	0.5942	0.5552	0.5166	0.4789	0.4426	0.4080	0.3754	0.3449	0.3166	0.2905	0.2665	0.2445	0.2245	0.2063	0.1896
8	0.5682	0.5243	0.4811	0.4393	0.3995	0.3621	0.3275	0.2957	0.2668	0.2406	0.2170	0.1959	0.1769	0.1600	0.1448
9	0.5453	0.4968	0.4493	0.4038	0.3611	0.3217	0.2858	0.2535	0.2247	0.1991	0.1765	0.1567	0.1392	0.1239	0.1104
10	0.5249	0.4720	0.4205	0.3719	0.3268	0.2860	0.2495	0.2173	0.1891	0.1647	0.1435	0.1252	0.1095	0.0959	0.0841

P=0.25, K=0.75

YEAR	0	5	10	15	20	25	30	35	40	45	50	55	60	65	70
0	1.0000	1.0000	1.0000	1.0000	1.0000	1.0000	1.0000	1.0000	1.0000	1.0000	1.0000	1.0000	1.0000	1.0000	1.0000
1	0.9115	0.9078	0.9041	0.9004	0.8968	0.8933	0.8898	0.8863	0.8829	0.8795	0.8762	0.8729	0.8697	0.8665	0.8633
2	0.8451	0.8364	0.8276	0.8189	0.8103	0.8017	0.7932	0.7848	0.7764	0.7681	0.7599	0.7519	0.7439	0.7360	0.7282
3	0.7927	0.7786	0.7644	0.7503	0.7361	0.7221	0.7081	0.6944	0.6807	0.6673	0.6541	0.6411	0.6283	0.6158	0.6035
4	0.7500	0.7305	0.7108	0.6911	0.6715	0.6521	0.6329	0.6140	0.5955	0.5774	0.5598	0.5426	0.5260	0.5098	0.4942
5	0.7142	0.6894	0.6643	0.6393	0.6145	0.5901	0.5661	0.5428	0.5201	0.4982	0.4772	0.4570	0.4376	0.4191	0.4015
6	0.6837	0.6537	0.6235	0.5934	0.5637	0.5348	0.5067	0.4796	0.4537	0.4291	0.4055	0.3835	0.3626	0.3430	0.3246
7	0.6571	0.6222	0.5870	0.5522	0.5182	0.4852	0.4537	0.4237	0.3955	0.3689	0.3442	0.3211	0.2997	0.2799	0.2616
8	0.6338	0.5941	0.5543	0.5150	0.4770	0.4407	0.4064	0.3743	0.3445	0.3169	0.2916	0.2685	0.2473	0.2280	0.2104
9	0.6131	0.5688	0.5245	0.4812	0.4397	0.4006	0.3641	0.3306	0.2999	0.2721	0.2469	0.2242	0.2039	0.1855	0.1691
10	0.5946	0.5458	0.4973	0.4503	0.4058	0.3643	0.3263	0.2919	0.2610	0.2334	0.2089	0.1872	0.1579	0.1509	0.1358

YEAR* \ GROWTH*	0	5	10	15	20	25	30	35	40	45	50	55	60	65	70
0	1.0000	1.0000	1.0000	1.0000	1.0000	1.0000	1.0000	1.0000	1.0000	1.0000	1.0000	1.0000	1.0000	1.0000	1.0000
1	0.9307	0.9277	0.9248	0.9219	0.9190	0.9162	0.9134	0.9106	0.9079	0.9052	0.9026	0.8999	0.8973	0.8948	0.8922
2	0.8776	0.8706	0.8635	0.8565	0.8495	0.8425	0.8355	0.8286	0.8218	0.8150	0.8082	0.8015	0.7949	0.7884	0.7819
3	0.8351	0.8236	0.8119	0.8002	0.7885	0.7768	0.7652	0.7536	0.7421	0.7307	0.7194	0.7083	0.6973	0.6866	0.6759
4	0.8000	0.7838	0.7674	0.7508	0.7343	0.7177	0.7013	0.6850	0.6690	0.6531	0.6376	0.6224	0.6075	0.5930	0.5789
5	0.7702	0.7494	0.7282	0.7068	0.6854	0.6642	0.6432	0.6225	0.6023	0.5825	0.5633	0.5447	0.5268	0.5094	0.4927
6	0.7445	0.7191	0.6932	0.6671	0.6411	0.6154	0.5902	0.5656	0.5417	0.5187	0.4967	0.4755	0.4553	0.4361	0.4178
7	0.7220	0.6921	0.6616	0.6309	0.6005	0.5707	0.5417	0.5137	0.4870	0.4614	0.4372	0.4143	0.3928	0.3725	0.3534
8	0.7021	0.6677	0.6327	0.5977	0.5632	0.5297	0.4974	0.4666	0.4375	0.4101	0.3845	0.3506	0.3384	0.3177	0.2985
9	0.6842	0.6455	0.6062	0.5670	0.5287	0.4918	0.4568	0.4238	0.3929	0.3643	0.3379	0.3136	0.2913	0.2707	0.2519
10	0.6681	0.6252	0.5817	0.5386	0.4968	0.4569	0.4195	0.3848	0.3528	0.3235	0.2968	0.2726	0.2506	0.2306	0.2125

P=0.25 *K=0.80*

YEAR*	0	5	10	15	20	25	30	35	40	45	50	55	60	65	70
0	1.0000	1.0000	1.0000	1.0000	1.0000	1.0000	1.0000	1.0000	1.0000	1.0000	1.0000	1.0000	1.0000	1.0000	1.0000
1	0.9490	0.9468	0.9446	0.9425	0.9403	0.9382	0.9361	0.9341	0.9321	0.9300	0.9281	0.9261	0.9241	0.9222	0.9203
2	0.9093	0.9040	0.8986	0.8933	0.8880	0.8826	0.8773	0.8720	0.8668	0.8615	0.8564	0.8512	0.8461	0.8410	0.8359
3	0.8770	0.8682	0.8592	0.8502	0.8411	0.8320	0.8229	0.8138	0.8047	0.7957	0.7868	0.7779	0.7691	0.7604	0.7518
4	0.8500	0.8374	0.8246	0.8116	0.7985	0.7854	0.7723	0.7592	0.7462	0.7333	0.7205	0.7080	0.6956	0.6835	0.6716
5	0.8268	0.8105	0.7937	0.7767	0.7595	0.7423	0.7251	0.7081	0.6912	0.6746	0.6584	0.6425	0.6270	0.6119	0.5972
6	0.8067	0.7865	0.7657	0.7446	0.7234	0.7021	0.6811	0.6603	0.6399	0.6200	0.6007	0.5819	0.5638	0.5464	0.5296
7	0.7888	0.7648	0.7401	0.7150	0.6898	0.6646	0.6399	0.6156	0.5921	0.5693	0.5474	0.5264	0.5063	0.4871	0.4688
8	0.7729	0.7451	0.7165	0.6874	0.6583	0.6295	0.6013	0.5740	0.5477	0.5225	0.4985	0.4757	0.4542	0.4338	0.4146
9	0.7585	0.7270	0.6945	0.6615	0.6287	0.5964	0.5651	0.5351	0.5065	0.4793	0.4538	0.4297	0.4072	0.3861	0.3664
10	0.7455	0.7103	0.6739	0.6372	0.6008	0.5653	0.5312	0.4988	0.4682	0.4396	0.4129	0.3880	0.3650	0.3435	0.3237

P=0.25 *K=0.85*

* GROWTH

YEAR*	0	5	10	15	20	25	30	35	40	45	50	55	60	65	70
0	1.0000	1.0000	1.0000	1.0000	1.0000	1.0000	1.0000	1.0000	1.0000	1.0000	1.0000	1.0000	1.0000	1.0000	1.0000
1	0.9667	0.9652	0.9637	0.9623	0.9609	0.9595	0.9581	0.9568	0.9554	0.9541	0.9527	0.9514	0.9501	0.9489	0.9476
2	0.9402	0.9366	0.9331	0.9295	0.9259	0.9223	0.9187	0.9151	0.9115	0.9079	0.9044	0.9008	0.8973	0.8938	0.8903
3	0.9185	0.9124	0.9063	0.9001	0.8939	0.8876	0.8813	0.8749	0.8686	0.8623	0.8560	0.8497	0.8435	0.8373	0.8312
4	0.9000	0.8913	0.8825	0.8734	0.8643	0.8550	0.8457	0.8364	0.8271	0.8178	0.8086	0.7994	0.7903	0.7814	0.7725
5	0.8840	0.8726	0.8609	0.8489	0.8367	0.8243	0.8119	0.7995	0.7871	0.7748	0.7626	0.7507	0.7389	0.7273	0.7159
6	0.8700	0.8558	0.8411	0.8260	0.8106	0.7951	0.7796	0.7641	0.7487	0.7335	0.7186	0.7040	0.6897	0.6758	0.6622
7	0.8575	0.8405	0.8228	0.8045	0.7860	0.7673	0.7487	0.7302	0.7119	0.6941	0.6766	0.6597	0.6432	0.6273	0.6119
8	0.8462	0.8264	0.8056	0.7843	0.7626	0.7408	0.7191	0.6977	0.6768	0.6565	0.6368	0.6178	0.5995	0.5819	0.5651
9	0.8360	0.8133	0.7895	0.7650	0.7402	0.7153	0.6907	0.6667	0.6434	0.6208	0.5991	0.5784	0.5585	0.5396	0.5216
10	0.8266	0.8011	0.7743	0.7466	0.7187	0.6909	0.6636	0.6370	0.6115	0.5870	0.5636	0.5413	0.5202	0.5002	0.4813

ρ=0.25
K=0.90

YEAR*	0	5	10	15	20	25	30	35	40	45	50	55	60	65	70
0	1.0000	1.0000	1.0000	1.0000	1.0000	1.0000	1.0000	1.0000	1.0000	1.0000	1.0000	1.0000	1.0000	1.0000	1.0000
1	0.9836	0.9829	0.9822	0.9815	0.9808	0.9801	0.9794	0.9787	0.9780	0.9774	0.9767	0.9761	0.9754	0.9748	0.9741
2	0.9704	0.9686	0.9668	0.9650	0.9632	0.9614	0.9595	0.9577	0.9559	0.9541	0.9522	0.9504	0.9486	0.9468	0.9450
3	0.9594	0.9564	0.9532	0.9501	0.9468	0.9436	0.9403	0.9370	0.9337	0.9304	0.9271	0.9238	0.9205	0.9172	0.9139
4	0.9500	0.9455	0.9409	0.9362	0.9315	0.9266	0.9217	0.9167	0.9117	0.9067	0.9017	0.8967	0.8918	0.8868	0.8819
5	0.9418	0.9358	0.9297	0.9233	0.9168	0.9102	0.9035	0.8968	0.8900	0.8832	0.8764	0.8597	0.8630	0.8564	0.8498
6	0.9344	0.9270	0.9192	0.9111	0.9029	0.8944	0.8858	0.8772	0.8686	0.8600	0.8514	0.8429	0.8346	0.8263	0.8182
7	0.9279	0.9189	0.9094	0.8995	0.8894	0.8790	0.8686	0.8580	0.8475	0.8371	0.8268	0.8167	0.8067	0.7969	0.7873
8	0.9219	0.9113	0.9001	0.8884	0.8764	0.8641	0.8517	0.8393	0.8269	0.8147	0.8028	0.7910	0.7795	0.7683	0.7574
9	0.9165	0.9043	0.8913	0.8777	0.8637	0.8495	0.8352	0.8209	0.8068	0.7929	0.7793	0.7560	0.7531	0.7406	0.7284
10	0.9115	0.8977	0.8829	0.8674	0.8514	0.8352	0.8190	0.8029	0.7870	0.7715	0.7564	0.7417	0.7275	0.7138	0.7005

ρ=0.25
K=0.95

P=0.30 K=0.70

YEAR \ GROWTH	0	5	10	15	20	25	30	35	40	45	50	55	60	65	70
0	1.0000	1.0000	1.0000	1.0000	1.0000	1.0000	1.0000	1.0000	1.0000	1.0000	1.0000	1.0000	1.0000	1.0000	1.0000
1	0.8737	0.8686	0.8635	0.8585	0.8537	0.8489	0.8441	0.8395	0.8349	0.8304	0.8260	0.8216	0.8173	0.8131	0.8089
2	0.7852	0.7739	0.7627	0.7516	0.7407	0.7299	0.7193	0.7089	0.6986	0.6884	0.6735	0.6687	0.6591	0.6497	0.6405
3	0.7187	0.7013	0.6839	0.6668	0.6499	0.6333	0.6170	0.6010	0.5854	0.5701	0.5552	0.5407	0.5267	0.5130	0.4997
4	0.6665	0.6432	0.6201	0.5973	0.5749	0.5530	0.5317	0.5110	0.4910	0.4717	0.4532	0.4353	0.4182	0.4018	0.3861
5	0.6241	0.5952	0.5663	0.5388	0.5116	0.4852	0.4598	0.4354	0.4122	0.3901	0.3692	0.3494	0.3307	0.3131	0.2965
6	0.5887	0.5548	0.5213	0.4887	0.4573	0.4272	0.3986	0.3716	0.3462	0.3225	0.3004	0.2799	0.2608	0.2432	0.2269
7	0.5587	0.5199	0.4820	0.4453	0.4102	0.3771	0.3462	0.3174	0.2909	0.2665	0.2442	0.2239	0.2054	0.1885	0.1732
8	0.5327	0.4895	0.4474	0.4071	0.3690	0.3337	0.3011	0.2714	0.2445	0.2202	0.1985	0.1790	0.1615	0.1460	0.1321
9	0.5101	0.4626	0.4167	0.3732	0.3327	0.2957	0.2622	0.2322	0.2055	0.1820	0.1612	0.1430	0.1270	0.1130	0.1007
10	0.4900	0.4386	0.3892	0.3429	0.3005	0.2623	0.2285	0.1987	0.1728	0.1503	0.1309	0.1142	0.0998	0.0874	0.0767

P=0.30 K=0.75

YEAR \ GROWTH	0	5	10	15	20	25	30	35	40	45	50	55	60	65	70
0	1.0000	1.0000	1.0000	1.0000	1.0000	1.0000	1.0000	1.0000	1.0000	1.0000	1.0000	1.0000	1.0000	1.0000	1.0000
1	0.8963	0.8926	0.8884	0.8843	0.8802	0.8762	0.8723	0.8684	0.8646	0.8608	0.8571	0.8534	0.8498	0.8463	0.8428
2	0.8228	0.8132	0.8037	0.7943	0.7850	0.7758	0.7666	0.7576	0.7488	0.7400	0.7314	0.7229	0.7145	0.7062	0.6981
3	0.7661	0.7511	0.7361	0.7212	0.7064	0.6918	0.6774	0.6632	0.6493	0.6356	0.6222	0.6090	0.5962	0.5837	0.5714
4	0.7209	0.7005	0.6801	0.6599	0.6399	0.6201	0.6008	0.5819	0.5635	0.5455	0.5281	0.5113	0.4950	0.4793	0.4641
5	0.6837	0.6581	0.6326	0.6073	0.5824	0.5580	0.5343	0.5114	0.4893	0.4680	0.4477	0.4282	0.4096	0.3919	0.3751
6	0.6522	0.6217	0.5913	0.5613	0.5320	0.5036	0.4762	0.4500	0.4251	0.4014	0.3791	0.3580	0.3383	0.3197	0.3023
7	0.6253	0.5900	0.5550	0.5207	0.4874	0.4554	0.4250	0.3963	0.3694	0.3442	0.3208	0.2991	0.2790	0.2604	0.2432
8	0.6018	0.5620	0.5227	0.4843	0.4475	0.4126	0.3798	0.3493	0.3211	0.2951	0.2714	0.2496	0.2298	0.2118	0.1954
9	0.5810	0.5370	0.4935	0.4516	0.4116	0.3743	0.3397	0.3080	0.2791	0.2530	0.2295	0.2083	0.1893	0.1722	0.1569
10	0.5625	0.5144	0.4671	0.4218	0.3792	0.3398	0.3040	0.2716	0.2427	0.2169	0.1940	0.1737	0.1558	0.1400	0.1260

* GROWTH YEAR*	0	5	10	15	20	25	30	35	40	45	50	55	60	65	70
0	1.0000	1.0000	1.0000	1.0000	1.0000	1.0000	1.0000	1.0000	1.0000	1.0000	1.0000	1.0000	1.0000	1.0000	1.0000
1	0.9190	0.9156	0.9123	0.9090	0.9058	0.9026	0.8994	0.8963	0.8933	0.8902	0.8873	0.8843	0.8814	0.8786	0.8758
2	0.8596	0.8518	0.8441	0.8364	0.8288	0.8212	0.8137	0.8063	0.7990	0.7917	0.7845	0.7775	0.7705	0.7635	0.7567
3	0.8133	0.8009	0.7885	0.7761	0.7637	0.7514	0.7392	0.7272	0.7153	0.7036	0.6920	0.6807	0.6695	0.6586	0.6479
4	0.7758	0.7587	0.7415	0.7244	0.7073	0.6903	0.6736	0.6570	0.6408	0.6250	0.6095	0.5943	0.5796	0.5653	0.5513
5	0.7445	0.7228	0.7010	0.6792	0.6575	0.6361	0.6150	0.5944	0.5744	0.5549	0.5361	0.5179	0.5004	0.4836	0.4674
6	0.7179	0.6917	0.6653	0.6390	0.6129	0.5874	0.5624	0.5383	0.5150	0.4926	0.4712	0.4508	0.4314	0.4129	0.3954
7	0.6947	0.6642	0.6334	0.6028	0.5727	0.5433	0.5150	0.4878	0.4619	0.4373	0.4140	0.3921	0.3715	0.3521	0.3340
8	0.6744	0.6396	0.6046	0.5699	0.5360	0.5032	0.4719	0.4422	0.4143	0.3881	0.3636	0.3408	0.3197	0.3000	0.2819
9	0.6563	0.6174	0.5783	0.5397	0.5023	0.4666	0.4328	0.4011	0.3716	0.3444	0.3192	0.2952	0.2750	0.2556	0.2378
10	0.6400	0.5971	0.5541	0.5119	0.4713	0.4329	0.3971	0.3639	0.3334	0.3056	0.2803	0.2573	0.2365	0.2176	0.2005

P=0.30 *K=0.80*

YEAR	0	5	10	15	20	25	30	35	40	45	50	55	60	65	70
0	1.0000	1.0000	1.0000	1.0000	1.0000	1.0000	1.0000	1.0000	1.0000	1.0000	1.0000	1.0000	1.0000	1.0000	1.0000
1	0.9403	0.9378	0.9353	0.9329	0.9304	0.9281	0.9257	0.9234	0.9211	0.9188	0.9166	0.9144	0.9122	0.9100	0.9079
2	0.8957	0.8898	0.8839	0.8780	0.8722	0.8664	0.8606	0.8549	0.8492	0.8436	0.8380	0.8325	0.8270	0.8216	0.8163
3	0.8603	0.8507	0.8411	0.8314	0.8217	0.8121	0.8025	0.7929	0.7835	0.7741	0.7648	0.7557	0.7466	0.7377	0.7290
4	0.8312	0.8178	0.8043	0.7907	0.7771	0.7634	0.7499	0.7365	0.7232	0.7101	0.6972	0.6846	0.6722	0.6600	0.6481
5	0.8067	0.7895	0.7720	0.7544	0.7368	0.7193	0.7018	0.6847	0.6678	0.6512	0.6351	0.6193	0.6040	0.5891	0.5747
6	0.7855	0.7645	0.7432	0.7217	0.7001	0.6787	0.6576	0.6369	0.6167	0.5971	0.5781	0.5598	0.5421	0.5251	0.5087
7	0.7670	0.7423	0.7171	0.6917	0.6663	0.6413	0.6167	0.5928	0.5697	0.5475	0.5261	0.5056	0.4861	0.4676	0.4499
8	0.7506	0.7222	0.6932	0.6640	0.6349	0.6065	0.5787	0.5520	0.5263	0.5019	0.4786	0.4566	0.4358	0.4161	0.3976
9	0.7358	0.7038	0.6711	0.6382	0.6057	0.5740	0.5434	0.5141	0.4863	0.4601	0.4354	0.4122	0.3905	0.3702	0.3513
10	0.7225	0.6869	0.6505	0.6140	0.5782	0.5435	0.5103	0.4789	0.4493	0.4217	0.3960	0.3721	0.3499	0.3293	0.3103

P=0.30 *K=0.85*

*
*
* GROWTH
*

YEAR*	0	5	10	15	20	25	30	35	40	45	50	55	60	65	70
0	1.0000	1.0000	1.0000	1.0000	1.0000	1.0000	1.0000	1.0000	1.0000	1.0000	1.0000	1.0000	1.0000	1.0000	1.0000
1	0.9609	0.9592	0.9576	0.9559	0.9543	0.9527	0.9512	0.9496	0.9481	0.9466	0.9451	0.9436	0.9421	0.9407	0.9393
2	0.9311	0.9271	0.9231	0.9191	0.9151	0.9112	0.9073	0.9033	0.8995	0.8956	0.8917	0.8879	0.8841	0.8804	0.8767
3	0.9070	0.9005	0.8938	0.8872	0.8805	0.8738	0.8671	0.8604	0.8537	0.8471	0.8405	0.8339	0.8274	0.8210	0.8147
4	0.8871	0.8778	0.8683	0.8588	0.8491	0.8395	0.8298	0.8201	0.8105	0.8010	0.7915	0.7822	0.7730	0.7639	0.7549
5	0.8700	0.8579	0.8455	0.8330	0.8204	0.8076	0.7949	0.7822	0.7697	0.7573	0.7450	0.7330	0.7212	0.7096	0.6983
6	0.8551	0.8403	0.8250	0.8094	0.7936	0.7778	0.7621	0.7464	0.7310	0.7159	0.7010	0.6865	0.6723	0.6586	0.6452
7	0.8420	0.8243	0.8060	0.7874	0.7686	0.7497	0.7310	0.7125	0.6944	0.6767	0.6594	0.6427	0.6265	0.6109	0.5958
8	0.8303	0.8098	0.7885	0.7668	0.7449	0.7231	0.7015	0.6803	0.6596	0.6396	0.6202	0.6016	0.5836	0.5664	0.5499
9	0.8197	0.7963	0.7721	0.7474	0.7225	0.6977	0.6734	0.6496	0.6266	0.6045	0.5833	0.5630	0.5436	0.5251	0.5075
10	0.8100	0.7839	0.7567	0.7289	0.7011	0.6735	0.6465	0.6204	0.5953	0.5713	0.5485	0.5268	0.5062	0.4867	0.4683

P=0.30
K=0.90

YEAR	0	5	10	15	20	25	30	35	40	45	50	55	60	65	70
0	1.0000	1.0000	1.0000	1.0000	1.0000	1.0000	1.0000	1.0000	1.0000	1.0000	1.0000	1.0000	1.0000	1.0000	1.0000
1	0.9808	0.9799	0.9791	0.9783	0.9775	0.9767	0.9759	0.9752	0.9744	0.9736	0.9729	0.9721	0.9714	0.9707	0.9700
2	0.9658	0.9638	0.9618	0.9598	0.9578	0.9557	0.9537	0.9517	0.9497	0.9477	0.9457	0.9438	0.9418	0.9399	0.9379
3	0.9536	0.9502	0.9468	0.9434	0.9399	0.9364	0.9329	0.9294	0.9259	0.9224	0.9189	0.9154	0.9119	0.9085	0.9050
4	0.9433	0.9385	0.9336	0.9286	0.9235	0.9183	0.9132	0.9080	0.9028	0.8976	0.8924	0.8873	0.8822	0.8771	0.8721
5	0.9344	0.9281	0.9216	0.9149	0.9081	0.9012	0.8943	0.8873	0.8803	0.8734	0.8665	0.8597	0.8529	0.8462	0.8396
6	0.9266	0.9188	0.9106	0.9022	0.8936	0.8849	0.8761	0.8673	0.8585	0.8498	0.8412	0.8327	0.8243	0.8160	0.8079
7	0.9197	0.9102	0.9004	0.8902	0.8797	0.8692	0.8585	0.8479	0.8373	0.8268	0.8165	0.8064	0.7964	0.7867	0.7772
8	0.9134	0.9024	0.8908	0.8787	0.8664	0.8540	0.8415	0.8290	0.8166	0.8045	0.7925	0.7808	0.7694	0.7583	0.7474
9	0.9077	0.8951	0.8817	0.8678	0.8536	0.8393	0.8249	0.8106	0.7965	0.7827	0.7692	0.7560	0.7432	0.7308	0.7188
10	0.9025	0.8882	0.8731	0.8573	0.8412	0.8249	0.8087	0.7926	0.7769	0.7615	0.7465	0.7319	0.7179	0.7043	0.6912

P=0.30
K=0.95

GROWTH

YEAR	0	5	10	15	20	25	30	35	40	45	50	55	60	65	70
0	1.0000	1.0000	1.0000	1.0000	1.0000	1.0000	1.0000	1.0000	1.0000	1.0000	1.0000	1.0000	1.0000	1.0000	1.0000
1	0.8569	0.8512	0.8457	0.8402	0.8349	0.8297	0.8245	0.8195	0.8145	0.8096	0.8048	0.8001	0.7955	0.7909	0.7864
2	0.7611	0.7490	0.7372	0.7256	0.7141	0.7028	0.6918	0.6809	0.6703	0.6598	0.6496	0.6396	0.6298	0.6202	0.6108
3	0.6912	0.6731	0.6552	0.6376	0.6204	0.6035	0.5870	0.5710	0.5553	0.5401	0.5253	0.5110	0.4971	0.4837	0.4706
4	0.6373	0.6135	0.5901	0.5672	0.5449	0.5231	0.5021	0.4818	0.4622	0.4434	0.4254	0.4081	0.3916	0.3759	0.3608
5	0.5942	0.5652	0.5368	0.5091	0.4823	0.4565	0.4318	0.4082	0.3859	0.3647	0.3447	0.3259	0.3081	0.2915	0.2758
6	0.5587	0.5249	0.4919	0.4600	0.4294	0.4003	0.3728	0.3470	0.3229	0.3004	0.2795	0.2601	0.2422	0.2257	0.2105
7	0.5288	0.4905	0.4533	0.4177	0.3840	0.3523	0.3228	0.2956	0.2705	0.2476	0.2267	0.2077	0.1904	0.1747	0.1604
8	0.5031	0.4607	0.4198	0.3809	0.3446	0.3109	0.2802	0.2522	0.2269	0.2042	0.1839	0.1658	0.1495	0.1351	0.1222
9	0.4808	0.4345	0.3902	0.3485	0.3100	0.2750	0.2435	0.2154	0.1905	0.1685	0.1492	0.1323	0.1175	0.1045	0.0931
10	0.4612	0.4113	0.3638	0.3197	0.2796	0.2437	0.2119	0.1842	0.1600	0.1391	0.1211	0.1056	0.0922	0.0807	0.0708

P=0.35
K=0.70

YEAR	0	5	10	15	20	25	30	35	40	45	50	55	60	65	70
0	1.0000	1.0000	1.0000	1.0000	1.0000	1.0000	1.0000	1.0000	1.0000	1.0000	1.0000	1.0000	1.0000	1.0000	1.0000
1	0.8829	0.8782	0.8736	0.8690	0.8646	0.8602	0.8559	0.8516	0.8475	0.8434	0.8393	0.8354	0.8315	0.8276	0.8238
2	0.8023	0.7921	0.7820	0.7720	0.7622	0.7524	0.7429	0.7335	0.7242	0.7151	0.7061	0.6974	0.6887	0.6802	0.6719
3	0.7424	0.7266	0.7110	0.6956	0.6804	0.6655	0.6508	0.6364	0.6222	0.6085	0.5950	0.5819	0.5691	0.5566	0.5445
4	0.6953	0.6744	0.6535	0.6330	0.6128	0.5930	0.5737	0.5549	0.5366	0.5189	0.5018	0.4854	0.4695	0.4542	0.4395
5	0.6571	0.6312	0.6054	0.5801	0.5553	0.5313	0.5079	0.4855	0.4639	0.4433	0.4236	0.4048	0.3869	0.3700	0.3539
6	0.6253	0.5946	0.5642	0.5345	0.5057	0.4778	0.4512	0.4258	0.4018	0.3790	0.3576	0.3375	0.3187	0.3010	0.2845
7	0.5981	0.5630	0.5283	0.4946	0.4621	0.4311	0.4017	0.3742	0.3484	0.3244	0.3021	0.2814	0.2624	0.2448	0.2286
8	0.5746	0.5352	0.4965	0.4591	0.4234	0.3898	0.3583	0.3292	0.3023	0.2777	0.2552	0.2347	0.2160	0.1990	0.1835
9	0.5540	0.5105	0.4681	0.4273	0.3889	0.3530	0.3200	0.2899	0.2625	0.2378	0.2156	0.1956	0.1777	0.1617	0.1473
10	0.5357	0.4884	0.4424	0.3986	0.3577	0.3202	0.2861	0.2555	0.2281	0.2037	0.1822	0.1631	0.1463	0.1314	0.1182

P=0.35
K=0.75

*
*
* GROWTH
*

P=0.35, K=0.80

YEAR*	0	5	10	15	20	25	30	35	40	45	50	55	60	65	70
0	1.0000	1.0000	1.0000	1.0000	1.0000	1.0000	1.0000	1.0000	1.0000	1.0000	1.0000	1.0000	1.0000	1.0000	1.0000
1	0.9079	0.9042	0.9005	0.8968	0.8933	0.8897	0.8863	0.8829	0.8795	0.8762	0.8730	0.8698	0.8666	0.8635	0.8605
2	0.8430	0.8346	0.8263	0.8181	0.8100	0.8020	0.7941	0.7863	0.7786	0.7710	0.7635	0.7561	0.7488	0.7417	0.7346
3	0.7937	0.7805	0.7676	0.7546	0.7418	0.7291	0.7166	0.7043	0.6921	0.6802	0.6685	0.6570	0.6458	0.6348	0.6241
4	0.7544	0.7367	0.7190	0.7014	0.6839	0.6667	0.6498	0.6332	0.6170	0.6012	0.5858	0.5708	0.5563	0.5422	0.5285
5	0.7220	0.6998	0.6776	0.6555	0.6337	0.6122	0.5913	0.5709	0.5512	0.5320	0.5136	0.4958	0.4788	0.4624	0.4467
6	0.6947	0.6681	0.6415	0.6152	0.5892	0.5639	0.5394	0.5157	0.4930	0.4712	0.4504	0.4307	0.4119	0.3941	0.3772
7	0.6712	0.6404	0.6096	0.5792	0.5495	0.5206	0.4929	0.4665	0.4414	0.4176	0.3951	0.3740	0.3542	0.3357	0.3183
8	0.6507	0.6158	0.5810	0.5467	0.5135	0.4815	0.4511	0.4224	0.3954	0.3702	0.3467	0.3248	0.3046	0.2858	0.2685
9	0.6325	0.5936	0.5550	0.5171	0.4806	0.4459	0.4132	0.3827	0.3544	0.3283	0.3042	0.2821	0.2619	0.2433	0.2264
10	0.6162	0.5735	0.5312	0.4900	0.4505	0.4134	0.3788	0.3470	0.3178	0.2911	0.2669	0.2450	0.2251	0.2071	0.1909

P=0.35, K=0.85

YEAR*	0	5	10	15	20	25	30	35	40	45	50	55	60	65	70
0	1.0000	1.0000	1.0000	1.0000	1.0000	1.0000	1.0000	1.0000	1.0000	1.0000	1.0000	1.0000	1.0000	1.0000	1.0000
1	0.9321	0.9292	0.9265	0.9238	0.9211	0.9184	0.9158	0.9133	0.9107	0.9082	0.9058	0.9034	0.9010	0.8986	0.8963
2	0.8830	0.8766	0.8703	0.8640	0.8578	0.8516	0.8454	0.8394	0.8334	0.8274	0.8216	0.8158	0.8100	0.8044	0.7988
3	0.8451	0.8349	0.8248	0.8146	0.8045	0.7945	0.7845	0.7746	0.7649	0.7553	0.7458	0.7364	0.7273	0.7182	0.7093
4	0.3144	0.8004	0.7864	0.7723	0.7583	0.7444	0.7306	0.7169	0.7035	0.6903	0.6774	0.6647	0.6524	0.6403	0.6285
5	0.7888	0.7711	0.7531	0.7352	0.7173	0.6995	0.6820	0.6648	0.6480	0.6315	0.6155	0.5999	0.5848	0.5702	0.5561
6	0.7670	0.7455	0.7238	0.7020	0.6803	0.6589	0.6379	0.6174	0.5974	0.5781	0.5594	0.5414	0.5241	0.5075	0.4916
7	0.7480	0.7228	0.6973	0.6718	0.6465	0.6217	0.5974	0.5739	0.5512	0.5294	0.5085	0.4886	0.4696	0.4516	0.4344
8	0.7312	0.7025	0.6733	0.6442	0.6154	0.5873	0.5600	0.5338	0.5088	0.4849	0.4623	0.4409	0.4207	0.4017	0.3837
9	0.7163	0.6840	0.6513	0.6186	0.5865	0.5553	0.5254	0.4968	0.4698	0.4443	0.4203	0.3979	0.3769	0.3572	0.3389
10	0.7028	0.6671	0.6308	0.5947	0.5595	0.5255	0.4931	0.4626	0.4339	0.4071	0.3822	0.3590	0.3376	0.3177	0.2993

*
*
* GROWTH
*

$P = 0.35$, $K = 0.90$

YEAR*	0	5	10	15	20	25	30	35	40	45	50	55	60	65	70
0	1.0000	1.0000	1.0000	1.0000	1.0000	1.0000	1.0000	1.0000	1.0000	1.0000	1.0000	1.0000	1.0000	1.0000	1.0000
1	0.9554	0.9535	0.9517	0.9499	0.9481	0.9463	0.9446	0.9429	0.9412	0.9395	0.9379	0.9362	0.9346	0.9331	0.9315
2	0.9225	0.9182	0.9139	0.9096	0.9053	0.9011	0.8969	0.8927	0.8885	0.8844	0.8804	0.8763	0.8723	0.8684	0.8645
3	0.8966	0.8896	0.8826	0.8755	0.8685	0.8614	0.8544	0.8474	0.8405	0.8336	0.8268	0.8201	0.8135	0.8069	0.8004
4	0.8754	0.8656	0.8557	0.8458	0.8358	0.8258	0.8158	0.8060	0.7961	0.7864	0.7769	0.7674	0.7581	0.7490	0.7400
5	0.8575	0.8449	0.8321	0.8192	0.8062	0.7932	0.7803	0.7675	0.7548	0.7423	0.7301	0.7180	0.7063	0.6948	0.6835
6	0.8420	0.8266	0.8109	0.7950	0.7790	0.7630	0.7472	0.7315	0.7161	0.7010	0.6862	0.6718	0.6578	0.6442	0.6311
7	0.8284	0.8102	0.7915	0.7727	0.7537	0.7348	0.7161	0.6977	0.6796	0.6621	0.6451	0.6286	0.6126	0.5972	0.5824
8	0.8163	0.7954	0.7738	0.7519	0.7300	0.7082	0.6867	0.6657	0.6453	0.6255	0.6064	0.5881	0.5705	0.5536	0.5374
9	0.8055	0.7818	0.7573	0.7324	0.7076	0.6830	0.6588	0.6354	0.6128	0.5910	0.5701	0.5502	0.5312	0.5131	0.4959
10	0.7956	0.7691	0.7418	0.7140	0.6863	0.6590	0.6324	0.6066	0.5820	0.5584	0.5360	0.5147	0.4946	0.4755	0.4575

$P = 0.35$, $K = 0.95$

YEAR*	0	5	10	15	20	25	30	35	40	45	50	55	60	65	70
0	1.0000	1.0000	1.0000	1.0000	1.0000	1.0000	1.0000	1.0000	1.0000	1.0000	1.0000	1.0000	1.0000	1.0000	1.0000
1	0.9780	0.9771	0.9762	0.9753	0.9744	0.9735	0.9726	0.9718	0.9709	0.9701	0.9693	0.9684	0.9676	0.9668	0.9660
2	0.9615	0.9593	0.9571	0.9549	0.9527	0.9506	0.9484	0.9462	0.9441	0.9420	0.9398	0.9378	0.9357	0.9336	0.9316
3	0.9483	0.9447	0.9410	0.9373	0.9337	0.9300	0.9263	0.9226	0.9189	0.9152	0.9116	0.9080	0.9044	0.9008	0.8973
4	0.9373	0.9322	0.9270	0.9217	0.9164	0.9110	0.9057	0.9003	0.8949	0.8896	0.8843	0.8791	0.8739	0.8687	0.8637
5	0.9279	0.9212	0.9144	0.9075	0.9004	0.8933	0.8862	0.8791	0.8720	0.8650	0.8580	0.8511	0.8443	0.8375	0.8309
6	0.9197	0.9115	0.9030	0.8943	0.8855	0.8766	0.8677	0.8588	0.8499	0.8412	0.8325	0.8240	0.8156	0.8073	0.7992
7	0.9124	0.9026	0.8925	0.8820	0.8714	0.8607	0.8499	0.8392	0.8286	0.8181	0.8078	0.7977	0.7878	0.7781	0.7686
8	0.9059	0.8945	0.8826	0.8704	0.8579	0.8454	0.8328	0.8203	0.8079	0.7958	0.7839	0.7722	0.7609	0.7499	0.7391
9	0.9000	0.8870	0.8734	0.8593	0.8450	0.8306	0.8162	0.8019	0.7878	0.7741	0.7607	0.7476	0.7349	0.7226	0.7107
10	0.8947	0.8800	0.8647	0.8487	0.8325	0.8162	0.8000	0.7840	0.7683	0.7530	0.7382	0.7238	0.7098	0.6964	0.6834

* GROWTH

YEAR*	0	5	10	15	20	25	30	35	40	45	50	55	60	65	70
0	1.0000	1.0000	1.0000	1.0000	1.0000	1.0000	1.0000	1.0000	1.0000	1.0000	1.0000	1.0000	1.0000	1.0000	1.0000
1	0.8410	0.8349	0.8289	0.8231	0.8173	0.8117	0.8062	0.8008	0.7955	0.7903	0.7852	0.7802	0.7753	0.7704	0.7657
2	0.7390	0.7264	0.7141	0.7020	0.6901	0.6785	0.6671	0.6560	0.6451	0.6345	0.6241	0.6139	0.6040	0.5943	0.5849
3	0.6665	0.6479	0.6297	0.6119	0.5945	0.5775	0.5610	0.5450	0.5294	0.5143	0.4997	0.4856	0.4720	0.4588	0.4461
4	0.6116	0.5876	0.5641	0.5412	0.5190	0.4976	0.4768	0.4569	0.4378	0.4195	0.4021	0.3854	0.3695	0.3543	0.3399
5	0.5632	0.5392	0.5110	0.4837	0.4574	0.4323	0.4083	0.3855	0.3639	0.3436	0.3245	0.3065	0.2896	0.2737	0.2589
6	0.5327	0.4993	0.4663	0.4357	0.4059	0.3778	0.3514	0.3266	0.3036	0.2822	0.2623	0.2440	0.2271	0.2115	0.1971
7	0.5031	0.4655	0.4292	0.3947	0.3621	0.3317	0.3036	0.2776	0.2539	0.2322	0.2124	0.1945	0.1782	0.1634	0.1501
8	0.4779	0.4364	0.3966	0.3592	0.3243	0.2922	0.2630	0.2365	0.2126	0.1912	0.1721	0.1550	0.1398	0.1263	0.1142
9	0.4560	0.4109	0.3681	0.3281	0.2914	0.2581	0.2283	0.2018	0.1783	0.1577	0.1395	0.1237	0.1098	0.0976	0.0869
10	0.4368	0.3884	0.3427	0.3006	0.2624	0.2284	0.1985	0.1723	0.1497	0.1301	0.1132	0.0987	0.0862	0.0754	0.0662

P=0.40
K=0.70

YEAR	0	5	10	15	20	25	30	35	40	45	50	55	60	65	70
0	1.0000	1.0000	1.0000	1.0000	1.0000	1.0000	1.0000	1.0000	1.0000	1.0000	1.0000	1.0000	1.0000	1.0000	1.0000
1	0.8697	0.3646	0.8596	0.8546	0.8498	0.8451	0.8405	0.8359	0.8315	0.8271	0.8228	0.8185	0.8144	0.8103	0.8063
2	0.7835	0.7728	0.7622	0.7517	0.7415	0.7314	0.7215	0.7117	0.7022	0.6928	0.6837	0.6747	0.6659	0.6572	0.6488
3	0.7209	0.7047	0.6887	0.6729	0.6574	0.6422	0.6274	0.6129	0.5987	0.5849	0.5715	0.5584	0.5457	0.5334	0.5215
4	0.6726	0.6513	0.6302	0.6095	0.5892	0.5695	0.5503	0.5317	0.5137	0.4963	0.4795	0.4634	0.4479	0.4331	0.4188
5	0.6333	0.6077	0.5819	0.5567	0.5321	0.5084	0.4855	0.4636	0.4425	0.4225	0.4034	0.3852	0.3680	0.3517	0.3362
6	0.6018	0.5711	0.5409	0.5116	0.4833	0.4561	0.4302	0.4056	0.3823	0.3604	0.3398	0.3206	0.3025	0.2856	0.2699
7	0.5746	0.5397	0.5055	0.4724	0.4407	0.4107	0.3823	0.3557	0.3310	0.3079	0.2866	0.2569	0.2488	0.2320	0.2166
8	0.5512	0.5123	0.4743	0.4379	0.4032	0.3708	0.3405	0.3126	0.2869	0.2634	0.2419	0.2224	0.2046	0.1884	0.1738
9	0.5308	0.4881	0.4466	0.4070	0.3699	0.3354	0.3038	0.2750	0.2489	0.2254	0.2042	0.1853	0.1683	0.1531	0.1394
10	0.5127	0.4664	0.4216	0.3793	0.3399	0.3039	0.2714	0.2422	0.2161	0.1930	0.1725	0.1544	0.1385	0.1243	0.1119

P=0.40
K=0.75

* GROWTH

YEAR*	0	5	10	15	20	25	30	35	40	45	50	55	60	65	70
0	1.0000	1.0000	1.0000	1.0000	1.0000	1.0000	1.0000	1.0000	1.0000	1.0000	1.0000	1.0000	1.0000	1.0000	1.0000
1	0.8973	0.8933	0.8892	0.8853	0.8814	0.8776	0.8739	0.8702	0.8666	0.8631	0.8596	0.8562	0.8528	0.8495	0.8462
2	0.8276	0.8188	0.8100	0.8014	0.7929	0.7845	0.7763	0.7682	0.7602	0.7523	0.7445	0.7369	0.7295	0.7221	0.7149
3	0.7758	0.7622	0.7488	0.7354	0.7223	0.7093	0.6966	0.6840	0.6717	0.6597	0.6479	0.6364	0.6252	0.6142	0.6035
4	0.7352	0.7170	0.6990	0.6811	0.6635	0.6462	0.6292	0.6126	0.5965	0.5808	0.5655	0.5507	0.5364	0.5225	0.5091
5	0.7021	0.6795	0.6571	0.6349	0.6130	0.5917	0.5710	0.5508	0.5313	0.5126	0.4945	0.4772	0.4605	0.4446	0.4294
6	0.6744	0.6476	0.6209	0.5946	0.5689	0.5439	0.5198	0.4966	0.4744	0.4531	0.4330	0.4138	0.3956	0.3783	0.3620
7	0.6507	0.6198	0.5891	0.5590	0.5297	0.5014	0.4743	0.4486	0.4241	0.4011	0.3794	0.3590	0.3399	0.3220	0.3052
8	0.6300	0.5952	0.5607	0.5270	0.4944	0.4632	0.4336	0.4057	0.3796	0.3553	0.3326	0.3116	0.2921	0.2740	0.2573
9	0.6118	0.5733	0.5351	0.4980	0.4623	0.4286	0.3969	0.3674	0.3400	0.3148	0.2917	0.2705	0.2510	0.2332	0.2169
10	0.5956	0.5534	0.5117	0.4714	0.4330	0.3970	0.3636	0.3329	0.3047	0.2791	0.2559	0.2348	0.2157	0.1985	0.1829

P=0.40

K=0.80

YEAR*	0	5	10	15	20	25	30	35	40	45	50	55	60	65	70
0	1.0000	1.0000	1.0000	1.0000	1.0000	1.0000	1.0000	1.0000	1.0000	1.0000	1.0000	1.0000	1.0000	1.0000	1.0000
1	0.9241	0.9211	0.9181	0.9151	0.9122	0.9093	0.9065	0.9037	0.9010	0.8983	0.8957	0.8931	0.8905	0.8880	0.8855
2	0.8713	0.8645	0.8573	0.8511	0.8445	0.8380	0.8316	0.8252	0.8190	0.8128	0.8067	0.8007	0.7947	0.7889	0.7832
3	0.8312	0.8206	0.8100	0.7995	0.7890	0.7787	0.7685	0.7584	0.7484	0.7386	0.7290	0.7195	0.7103	0.7012	0.6922
4	0.7993	0.7848	0.7704	0.7560	0.7417	0.7276	0.7136	0.6999	0.6864	0.6732	0.6602	0.6476	0.6353	0.6233	0.6116
5	0.7729	0.7547	0.7365	0.7183	0.7002	0.6824	0.6649	0.6477	0.6309	0.6146	0.5988	0.5834	0.5685	0.5541	0.5402
6	0.7506	0.7287	0.7067	0.6848	0.6631	0.6418	0.6209	0.6006	0.5809	0.5619	0.5435	0.5259	0.5089	0.4927	0.4771
7	0.7312	0.7058	0.6802	0.6547	0.6295	0.6048	0.5809	0.5577	0.5354	0.5141	0.4937	0.4742	0.4557	0.4381	0.4214
8	0.7143	0.6853	0.6562	0.6272	0.5987	0.5709	0.5441	0.5184	0.4939	0.4706	0.4485	0.4277	0.4080	0.3895	0.3721
9	0.6992	0.6668	0.6342	0.6018	0.5701	0.5395	0.5101	0.4822	0.4558	0.4310	0.4077	0.3858	0.3654	0.3464	0.3286
10	0.6857	0.6499	0.6139	0.5783	0.5436	0.5103	0.4786	0.4488	0.4209	0.3948	0.3706	0.3481	0.3273	0.3080	0.2901

P=0.40

K=0.85

* GROWTH

P=0.40, K=0.90

YEAR*	0	5	10	15	20	25	30	35	40	45	50	55	60	65	70
0	1.0000	1.0000	1.0000	1.0000	1.0000	1.0000	1.0000	1.0000	1.0000	1.0000	1.0000	1.0000	1.0000	1.0000	1.0000
1	0.9501	0.9481	0.9461	0.9441	0.9421	0.9402	0.9383	0.9365	0.9346	0.9328	0.9311	0.9293	0.9276	0.9259	0.9242
2	0.9145	0.9099	0.9053	0.9008	0.8962	0.8917	0.8873	0.8829	0.8785	0.8742	0.8700	0.8658	0.8616	0.8575	0.8535
3	0.8871	0.8797	0.8723	0.8649	0.8576	0.8503	0.8430	0.8358	0.8287	0.8217	0.8147	0.8079	0.8011	0.7944	0.7878
4	0.8648	0.8547	0.8444	0.8342	0.8239	0.8137	0.8035	0.7935	0.7835	0.7737	0.7640	0.7545	0.7452	0.7360	0.7271
5	0.8462	0.8332	0.8201	0.8069	0.7937	0.7806	0.7675	0.7546	0.7419	0.7294	0.7171	0.7051	0.6934	0.6820	0.6709
6	0.8303	0.8145	0.7985	0.7824	0.7662	0.7501	0.7342	0.7186	0.7032	0.6882	0.6735	0.6592	0.6454	0.6320	0.6190
7	0.8163	0.7978	0.7789	0.7599	0.7408	0.7218	0.7032	0.6849	0.6670	0.6496	0.6328	0.6165	0.6008	0.5856	0.5710
8	0.8040	0.7827	0.7610	0.7390	0.7170	0.6953	0.6740	0.6532	0.6330	0.6135	0.5947	0.5766	0.5593	0.5427	0.5268
9	0.7930	0.7690	0.7444	0.7195	0.6947	0.6703	0.6464	0.6232	0.6009	0.5794	0.5589	0.5393	0.5207	0.5029	0.4860
10	0.7830	0.7563	0.7288	0.7011	0.6736	0.6465	0.6202	0.5949	0.5706	0.5474	0.5254	0.5045	0.4847	0.4660	0.4483

P=0.40, K=0.95

YEAR*	0	5	10	15	20	25	30	35	40	45	50	55	60	65	70
0	1.0000	1.0000	1.0000	1.0000	1.0000	1.0000	1.0000	1.0000	1.0000	1.0000	1.0000	1.0000	1.0000	1.0000	1.0000
1	0.9754	0.9744	0.9734	0.9724	0.9714	0.9704	0.9695	0.9686	0.9676	0.9667	0.9658	0.9649	0.9641	0.9632	0.9623
2	0.9574	0.9551	0.9527	0.9504	0.9481	0.9457	0.9435	0.9412	0.9389	0.9367	0.9344	0.9322	0.9301	0.9279	0.9258
3	0.9433	0.9395	0.9357	0.9318	0.9279	0.9241	0.9202	0.9164	0.9126	0.9088	0.9051	0.9013	0.8976	0.8940	0.8904
4	0.9317	0.9264	0.9210	0.9155	0.9100	0.9045	0.8990	0.8935	0.8880	0.8826	0.8772	0.8719	0.8666	0.8614	0.8563
5	0.9219	0.9150	0.9080	0.9008	0.8936	0.8864	0.8791	0.8719	0.8647	0.8576	0.8505	0.8436	0.8367	0.8300	0.8234
6	0.9134	0.9049	0.8962	0.8874	0.8784	0.8694	0.8604	0.8514	0.8425	0.8336	0.8250	0.8164	0.8080	0.7998	0.7917
7	0.9059	0.8959	0.8855	0.8749	0.8641	0.8533	0.8424	0.8317	0.8211	0.8106	0.8003	0.7902	0.7803	0.7707	0.7613
8	0.8992	0.8876	0.8755	0.8631	0.8505	0.8379	0.8252	0.8127	0.8004	0.7883	0.7764	0.7649	0.7536	0.7426	0.7320
9	0.8932	0.8799	0.8661	0.8519	0.8375	0.8230	0.8086	0.7944	0.7804	0.7667	0.7534	0.7404	0.7278	0.7156	0.7038
10	0.8877	0.8728	0.8573	0.8413	0.8250	0.8087	0.7925	0.7766	0.7610	0.7458	0.7310	0.7167	0.7029	0.6896	0.6767

*
*
* GROWTH
*

YEAR*	0	5	10	15	20	25	30	35	40	45	50	55	60	65	70
0	1.0000	1.0000	1.0000	1.0000	1.0000	1.0000	1.0000	1.0000	1.0000	1.0000	1.0000	1.0000	1.0000	1.0000	1.0000
1	0.8260	0.8195	0.8131	0.8069	0.8008	0.7948	0.7890	0.7833	0.7777	0.7722	0.7669	0.7616	0.7565	0.7514	0.7465
2	0.7187	0.7057	0.6930	0.6805	0.6684	0.6565	0.6449	0.6336	0.6225	0.6118	0.6013	0.5910	0.5811	0.5714	0.5619
3	0.6443	0.6254	0.6070	0.5890	0.5716	0.5546	0.5381	0.5222	0.5068	0.4919	0.4775	0.4536	0.4502	0.4373	0.4249
4	0.5887	0.5647	0.5412	0.5185	0.4965	0.4753	0.4550	0.4355	0.4169	0.3991	0.3822	0.3660	0.3507	0.3361	0.3222
5	0.5453	0.5165	0.4886	0.4618	0.4360	0.4115	0.3882	0.3661	0.3453	0.3257	0.3074	0.2901	0.2739	0.2588	0.2447
6	0.5101	0.4770	0.4452	0.4148	0.3859	0.3587	0.3332	0.3095	0.2874	0.2669	0.2480	0.2305	0.2144	0.1996	0.1860
7	0.4808	0.4439	0.4085	0.3750	0.3436	0.3143	0.2873	0.2625	0.2399	0.2192	0.2005	0.1835	0.1680	0.1541	0.1414
8	0.4560	0.4155	0.3769	0.3407	0.3072	0.2765	0.2486	0.2234	0.2007	0.1804	0.1623	0.1462	0.1318	0.1190	0.1076
9	0.4346	0.3907	0.3493	0.3108	0.2757	0.2439	0.2156	0.1904	0.1682	0.1486	0.1315	0.1165	0.1034	0.0919	0.0819
10	0.4159	0.3689	0.3249	0.2845	0.2481	0.2157	0.1873	0.1625	0.1411	0.1225	0.1066	0.0929	0.0811	0.0710	0.0623

P=0.45
K=0.70

YEAR*	0	5	10	15	20	25	30	35	40	45	50	55	60	65	70
0	1.0000	1.0000	1.0000	1.0000	1.0000	1.0000	1.0000	1.0000	1.0000	1.0000	1.0000	1.0000	1.0000	1.0000	1.0000
1	0.8571	0.8516	0.8463	0.8411	0.8359	0.8309	0.8260	0.8212	0.8165	0.8118	0.8073	0.8028	0.7984	0.7942	0.7899
2	0.7661	0.7549	0.7439	0.7331	0.7225	0.7122	0.7020	0.6920	0.6823	0.6728	0.6634	0.6543	0.6454	0.6367	0.6282
3	0.7014	0.6848	0.6685	0.6525	0.6369	0.6216	0.6067	0.5921	0.5780	0.5642	0.5509	0.5379	0.5254	0.5132	0.5014
4	0.6522	0.6307	0.6095	0.5887	0.5685	0.5489	0.5299	0.5115	0.4938	0.4767	0.4603	0.4446	0.4294	0.4150	0.4011
5	0.6131	0.5869	0.5612	0.5362	0.5120	0.4886	0.4661	0.4447	0.4242	0.4047	0.3861	0.3686	0.3519	0.3362	0.3213
6	0.5810	0.5505	0.5206	0.4917	0.4640	0.4374	0.4122	0.3883	0.3658	0.3446	0.3248	0.3062	0.2888	0.2726	0.2575
7	0.5540	0.5194	0.4857	0.4533	0.4224	0.3932	0.3657	0.3401	0.3162	0.2940	0.2736	0.2547	0.2373	0.2212	0.2065
8	0.5308	0.4924	0.4552	0.4196	0.3860	0.3546	0.3254	0.2985	0.2738	0.2513	0.2307	0.2120	0.1950	0.1796	0.1656
9	0.5106	0.4686	0.4281	0.3897	0.3537	0.3205	0.2901	0.2624	0.2374	0.2149	0.1947	0.1766	0.1604	0.1458	0.1328
10	0.4929	0.4474	0.4038	0.3628	0.3248	0.2902	0.2589	0.2309	0.2060	0.1839	0.1644	0.1471	0.1319	0.1184	0.1066

P=0.45
K=0.75

GROWTH

P=0.45, K=0.80

YEAR	0	5	10	15	20	25	30	35	40	45	50	55	60	65	70
0	1.0000	1.0000	1.0000	1.0000	1.0000	1.0000	1.0000	1.0000	1.0000	1.0000	1.0000	1.0000	1.0000	1.0000	1.0000
1	0.8873	0.8829	0.8786	0.8744	0.8702	0.8662	0.8622	0.8583	0.8545	0.8507	0.8470	0.8434	0.8398	0.8363	0.8328
2	0.8133	0.8041	0.7950	0.7860	0.7772	0.7685	0.7600	0.7516	0.7434	0.7353	0.7274	0.7196	0.7120	0.7046	0.6972
3	0.7595	0.7456	0.7317	0.7181	0.7047	0.6915	0.6786	0.6660	0.6536	0.6415	0.6297	0.6182	0.6070	0.5961	0.5854
4	0.7179	0.6994	0.6811	0.6630	0.6453	0.6280	0.6110	0.5945	0.5785	0.5629	0.5478	0.5332	0.5191	0.5055	0.4923
5	0.6842	0.6614	0.6389	0.6167	0.5949	0.5738	0.5532	0.5333	0.5142	0.4957	0.4780	0.4611	0.4448	0.4293	0.4145
6	0.6563	0.6294	0.6027	0.5766	0.5512	0.5266	0.5028	0.4801	0.4583	0.4376	0.4180	0.3993	0.3816	0.3649	0.3491
7	0.6325	0.6016	0.5712	0.5414	0.5125	0.4848	0.4583	0.4332	0.4094	0.3870	0.3659	0.3461	0.3276	0.3103	0.2941
8	0.6118	0.5772	0.5431	0.5099	0.4779	0.4474	0.4186	0.3915	0.3661	0.3425	0.3206	0.3003	0.2814	0.2640	0.2479
9	0.5937	0.5555	0.5179	0.4814	0.4466	0.4137	0.3829	0.3543	0.3278	0.3034	0.2811	0.2506	0.2418	0.2246	0.2089
10	0.5776	0.5359	0.4949	0.4555	0.4180	0.3830	0.3506	0.3209	0.2937	0.2689	0.2465	0.2262	0.2078	0.1911	0.1761

P=0.45, K=0.85

YEAR	0	5	10	15	20	25	30	35	40	45	50	55	60	65	70
0	1.0000	1.0000	1.0000	1.0000	1.0000	1.0000	1.0000	1.0000	1.0000	1.0000	1.0000	1.0000	1.0000	1.0000	1.0000
1	0.9166	0.9133	0.9100	0.9068	0.9037	0.9006	0.8976	0.8947	0.8918	0.8889	0.8861	0.8833	0.8806	0.8779	0.8753
2	0.8603	0.8532	0.8461	0.8392	0.8323	0.8255	0.8188	0.8122	0.8058	0.7994	0.7931	0.7869	0.7809	0.7749	0.7690
3	0.8185	0.8075	0.7965	0.7857	0.7750	0.7644	0.7540	0.7438	0.7337	0.7238	0.7140	0.7045	0.6952	0.6860	0.6771
4	0.7855	0.7707	0.7560	0.7413	0.7269	0.7126	0.6985	0.6847	0.6712	0.6580	0.6451	0.6326	0.6203	0.6084	0.5969
5	0.7585	0.7400	0.7216	0.7032	0.6851	0.6672	0.6497	0.6327	0.6160	0.5999	0.5842	0.5690	0.5543	0.5402	0.5265
6	0.7358	0.7137	0.6916	0.6697	0.6480	0.6268	0.6061	0.5860	0.5665	0.5478	0.5297	0.5124	0.4958	0.4799	0.4646
7	0.7163	0.6907	0.6650	0.6396	0.6146	0.5902	0.5665	0.5437	0.5218	0.5008	0.4808	0.4618	0.4437	0.4265	0.4102
8	0.6992	0.6702	0.6411	0.6123	0.5841	0.5567	0.5303	0.5051	0.4811	0.4583	0.4367	0.4163	0.3972	0.3791	0.3621
9	0.6841	0.6517	0.6192	0.5872	0.5559	0.5258	0.4970	0.4696	0.4438	0.4195	0.3968	0.3755	0.3556	0.3370	0.3197
10	0.6705	0.6349	0.5991	0.5639	0.5298	0.4971	0.4661	0.4370	0.4097	0.3842	0.3606	0.3387	0.3184	0.2996	0.2823

*
 *
  * GROWTH
   *

P=0.45
K=0.90

YEAR*	0	5	10	15	20	25	30	35	40	45	50	55	60	65	70
0	1.0000	1.0000	1.0000	1.0000	1.0000	1.0000	1.0000	1.0000	1.0000	1.0000	1.0000	1.0000	1.0000	1.0000	1.0000
1	0.9451	0.9429	0.9407	0.9386	0.9365	0.9344	0.9324	0.9304	0.9284	0.9265	0.9246	0.9227	0.9209	0.9191	0.9173
2	0.9070	0.9022	0.8973	0.8925	0.8878	0.8831	0.8785	0.8739	0.8694	0.8649	0.8605	0.8551	0.8518	0.8476	0.8434
3	0.8782	0.8705	0.8629	0.8553	0.8477	0.8402	0.8327	0.8254	0.8181	0.8109	0.8038	0.7969	0.7900	0.7832	0.7766
4	0.8551	0.8447	0.8341	0.8236	0.8132	0.8028	0.7925	0.7823	0.7722	0.7624	0.7527	0.7431	0.7338	0.7246	0.7156
5	0.8360	0.8227	0.8093	0.7959	0.7825	0.7693	0.7561	0.7432	0.7305	0.7180	0.7057	0.6938	0.6822	0.6708	0.6598
6	0.8197	0.8036	0.7874	0.7711	0.7548	0.7387	0.7228	0.7072	0.6919	0.6769	0.6624	0.6483	0.6345	0.6213	0.6084
7	0.8055	0.7867	0.7676	0.7485	0.7294	0.7104	0.6918	0.6737	0.6559	0.6387	0.6221	0.6060	0.5905	0.5755	0.5611
8	0.7930	0.7715	0.7495	0.7276	0.7057	0.6840	0.6629	0.6422	0.6223	0.6030	0.5844	0.5566	0.5495	0.5332	0.5176
9	0.7818	0.7576	0.7329	0.7081	0.6834	0.6592	0.6355	0.6126	0.5906	0.5694	0.5492	0.5299	0.5115	0.4941	0.4774
10	0.7717	0.7449	0.7174	0.6898	0.6624	0.6357	0.6097	0.5846	0.5607	0.5379	0.5162	0.4957	0.4762	0.4578	0.4404

P=0.45
K=0.95

YEAR*	0	5	10	15	20	25	30	35	40	45	50	55	60	65	70
0	1.0000	1.0000	1.0000	1.0000	1.0000	1.0000	1.0000	1.0000	1.0000	1.0000	1.0000	1.0000	1.0000	1.0000	1.0000
1	0.9729	0.9718	0.9707	0.9696	0.9686	0.9675	0.9665	0.9655	0.9645	0.9635	0.9625	0.9516	0.9607	0.9597	0.9588
2	0.9536	0.9511	0.9486	0.9462	0.9437	0.9413	0.9389	0.9365	0.9341	0.9318	0.9295	0.9272	0.9249	0.9227	0.9204
3	0.9387	0.9347	0.9307	0.9267	0.9227	0.9187	0.9147	0.9108	0.9069	0.9030	0.8991	0.8953	0.8916	0.8879	0.8842
4	0.9266	0.9211	0.9155	0.9099	0.9042	0.8986	0.8929	0.8873	0.8818	0.8763	0.8708	0.8654	0.8601	0.8549	0.8497
5	0.9165	0.9094	0.9021	0.8948	0.8875	0.8801	0.8728	0.8655	0.8582	0.8510	0.8440	0.8370	0.8301	0.8233	0.8167
6	0.9077	0.8990	0.8901	0.8811	0.8720	0.8629	0.8538	0.8448	0.8358	0.8270	0.8183	0.8097	0.8014	0.7932	0.7851
7	0.9000	0.8898	0.8792	0.8684	0.8576	0.8467	0.8358	0.8250	0.8144	0.8039	0.7937	0.7836	0.7738	0.7642	0.7548
8	0.8932	0.8813	0.8691	0.8566	0.8439	0.8312	0.8186	0.8061	0.7938	0.7817	0.7699	0.7584	0.7472	0.7363	0.7257
9	0.8871	0.8736	0.8596	0.8453	0.8309	0.8164	0.8020	0.7878	0.7738	0.7602	0.7470	0.7341	0.7216	0.7095	0.6977
10	0.8815	0.8664	0.8507	0.8346	0.8183	0.8020	0.7859	0.7700	0.7545	0.7394	0.7248	0.7106	0.6969	0.6836	0.6709

* GROWTH

P=0.50, K=0.70

YEAR	0	5	10	15	20	25	30	35	40	45	50	55	60	65	70
0	1.0000	1.0000	1.0000	1.0000	1.0000	1.0000	1.0000	1.0000	1.0000	1.0000	1.0000	1.0000	1.0000	1.0000	1.0000
1	0.3117	0.8048	0.7981	0.7916	0.7852	0.7789	0.7728	0.7669	0.7611	0.7554	0.7498	0.7443	0.7390	0.7338	0.7287
2	0.7000	0.6867	0.6736	0.6609	0.6485	0.6365	0.6247	0.6133	0.6021	0.5913	0.5808	0.5705	0.5605	0.5508	0.5414
3	0.6241	0.6050	0.5865	0.5685	0.5510	0.5341	0.5178	0.5020	0.4868	0.4721	0.4579	0.4443	0.4312	0.4186	0.4065
4	0.5682	0.5442	0.5208	0.4983	0.4766	0.4558	0.4359	0.4168	0.3987	0.3814	0.3649	0.3492	0.3344	0.3203	0.3069
5	0.5249	0.4964	0.4689	0.4425	0.4173	0.3934	0.3707	0.3494	0.3293	0.3104	0.2927	0.2761	0.2606	0.2461	0.2325
6	0.4900	0.4575	0.4263	0.3966	0.3685	0.3422	0.3176	0.2947	0.2734	0.2538	0.2357	0.2190	0.2036	0.1895	0.1765
7	0.4612	0.4250	0.3905	0.3580	0.3276	0.2994	0.2734	0.2497	0.2280	0.2082	0.1903	0.1741	0.1594	0.1461	0.1341
8	0.4368	0.3973	0.3598	0.3248	0.2925	0.2630	0.2363	0.2122	0.1905	0.1712	0.1540	0.1386	0.1250	0.1128	0.1020
9	0.4159	0.3732	0.3331	0.2960	0.2622	0.2318	0.2047	0.1807	0.15[illegible]5	0.1410	0.1247	0.1104	0.0980	0.0871	0.0776
10	0.3977	0.3521	0.3095	0.2707	0.2358	0.2049	0.1777	0.1542	0.1338	0.1162	0.1011	0.0881	0.0769	0.0673	0.0590

P=0.50, K=0.75

YEAR	0	5	10	15	20	25	30	35	40	45	50	55	60	65	70
0	1.0000	1.0000	1.0000	1.0000	1.0000	1.0000	1.0000	1.0000	1.0000	1.0000	1.0000	1.0000	1.0000	1.0000	1.0000
1	0.8451	0.8393	0.8337	0.8282	0.8228	0.8175	0.8123	0.8073	0.8023	0.7975	0.7927	0.7881	0.7935	0.7791	0.7747
2	0.7500	0.7384	0.7271	0.7160	0.7052	0.6946	0.6842	0.6741	0.6642	0.6546	0.6451	0.6359	0.6270	0.6182	0.6096
3	0.6837	0.6658	0.6503	0.6341	0.6184	0.6030	0.5881	0.5736	0.5595	0.5458	0.5326	0.5198	0.5074	0.4954	0.4838
4	0.6338	0.6121	0.5909	0.5702	0.5501	0.5306	0.5118	0.4937	0.4763	0.4595	0.4435	0.4281	0.4133	0.3992	0.3857
5	0.5946	0.5684	0.5429	0.5181	0.4942	0.4712	0.4492	0.4282	0.4082	0.3892	0.3712	0.3541	0.3380	0.3227	0.3083
6	0.5625	0.5322	0.5027	0.4743	0.4470	0.4211	0.3965	0.3732	0.3514	0.3309	0.3117	0.2938	0.2770	0.2614	0.2469
7	0.5357	0.5015	0.4684	0.4366	0.4065	0.3780	0.3514	0.3265	0.3034	0.2821	0.2623	0.2442	0.2274	0.2120	0.1978
8	0.5127	0.4749	0.4385	0.4037	0.3711	0.3406	0.3123	0.2864	0.2626	0.2409	0.2211	0.2031	0.1868	0.1720	0.1586
9	0.4929	0.4516	0.4120	0.3746	0.3397	0.3076	0.2782	0.2516	0.2275	0.2059	0.1865	0.1691	0.1536	0.1397	0.1272
10	0.4754	0.4309	0.3884	0.3435	0.3118	0.2784	0.2483	0.2213	0.1974	0.1762	0.1574	0.1409	0.1263	0.1134	0.1020

*
*
* GROWTH
*

P=0.50, K=0.80

YEAR*	0	5	10	15	20	25	30	35	40	45	50	55	60	65	70
0	1.0000	1.0000	1.0000	1.0000	1.0000	1.0000	1.0000	1.0000	1.0000	1.0000	1.0000	1.0000	1.0000	1.0000	1.0000
1	0.8776	0.8730	0.8684	0.8640	0.8596	0.8553	0.8511	0.8470	0.8430	0.8390	0.8351	0.8313	0.8276	0.8239	0.8203
2	0.8000	0.7904	0.7810	0.7718	0.7627	0.7538	0.7450	0.7365	0.7281	0.7198	0.7118	0.7039	0.6962	0.6886	0.6812
3	0.7445	0.7303	0.7162	0.7024	0.6888	0.6755	0.6625	0.6498	0.6373	0.6252	0.6135	0.6020	0.5908	0.5799	0.5694
4	0.7021	0.6834	0.6649	0.6468	0.6290	0.6117	0.5948	0.5784	0.5625	0.5471	0.5322	0.5178	0.5039	0.4905	0.4776
5	0.6681	0.6452	0.6226	0.6004	0.5788	0.5578	0.5375	0.5179	0.4991	0.4810	0.4636	0.4470	0.4311	0.4160	0.4015
6	0.6400	0.6131	0.5866	0.5607	0.5355	0.5112	0.4879	0.4656	0.4443	0.4241	0.4049	0.3867	0.3695	0.3532	0.3379
7	0.6162	0.5855	0.5553	0.5259	0.4974	0.4702	0.4443	0.4197	0.3965	0.3747	0.3542	0.3350	0.3170	0.3002	0.2845
8	0.5956	0.5613	0.5276	0.4949	0.4635	0.4337	0.4055	0.3791	0.3544	0.3315	0.3102	0.2905	0.2722	0.2553	0.2397
9	0.5776	0.5398	0.5027	0.4669	0.4328	0.4007	0.3707	0.3429	0.3172	0.2935	0.2719	0.2520	0.2338	0.2172	0.2020
10	0.5517	0.5205	0.4802	0.4415	0.4050	0.3709	0.3394	0.3104	0.2841	0.2601	0.2384	0.2187	0.2009	0.1848	0.1703

P=0.50, K=0.85

YEAR*	0	5	10	15	20	25	30	35	40	45	50	55	60	65	70
0	1.0000	1.0000	1.0000	1.0000	1.0000	1.0000	1.0000	1.0000	1.0000	1.0000	1.0000	1.0000	1.0000	1.0000	1.0000
1	0.9093	0.9058	0.9023	0.8990	0.8957	0.8924	0.8892	0.8861	0.8830	0.8800	0.8770	0.8741	0.8713	0.8684	0.8657
2	0.8500	0.8426	0.8353	0.8280	0.8209	0.8139	0.8071	0.8003	0.7936	0.7871	0.7807	0.7744	0.7682	0.7621	0.7561
3	0.8067	0.7954	0.7842	0.7731	0.7622	0.7515	0.7409	0.7305	0.7203	0.7103	0.7006	0.6910	0.6816	0.6725	0.6635
4	0.7729	0.7578	0.7429	0.7281	0.7135	0.6991	0.6850	0.6712	0.6577	0.6445	0.6317	0.6192	0.6071	0.5953	0.5838
5	0.7455	0.7268	0.7081	0.6897	0.6715	0.6537	0.6363	0.6193	0.6028	0.5868	0.5713	0.5563	0.5418	0.5279	0.5144
6	0.7225	0.7002	0.6781	0.6561	0.6346	0.6135	0.5930	0.5731	0.5539	0.5354	0.5176	0.5006	0.4843	0.4686	0.4537
7	0.7028	0.6771	0.6515	0.6262	0.6014	0.5772	0.5539	0.5314	0.5098	0.4892	0.4696	0.4509	0.4332	0.4163	0.4004
8	0.6857	0.6566	0.6277	0.5991	0.5712	0.5442	0.5182	0.4934	0.4698	0.4475	0.4263	0.4064	0.3876	0.3700	0.3534
9	0.6705	0.6382	0.6060	0.5743	0.5434	0.5137	0.4854	0.4586	0.4333	0.4095	0.3873	0.3665	0.3470	0.3289	0.3120
10	0.6570	0.6215	0.5861	0.5513	0.5177	0.4856	0.4552	0.4266	0.3999	0.3750	0.3519	0.3305	0.3107	0.2924	0.2754

* GROWTH

YEAR*	0	5	10	15	20	25	30	35	40	45	50	55	60	65	70
0	1.0000	1.0000	1.0000	1.0000	1.0000	1.0000	1.0000	1.0000	1.0000	1.0000	1.0000	1.0000	1.0000	1.0000	1.0000
1	0.9402	0.9379	0.9356	0.9333	0.9311	0.9289	0.9267	0.9246	0.9225	0.9205	0.9185	0.9165	0.9145	0.9126	0.9107
2	0.9000	0.8949	0.8898	0.8849	0.8799	0.8751	0.8703	0.8655	0.8608	0.8562	0.8517	0.8472	0.8428	0.8385	0.8342
3	0.8700	0.8621	0.8542	0.8463	0.8386	0.8309	0.8233	0.8158	0.8084	0.8011	0.7940	0.7869	0.7800	0.7732	0.7665
4	0.8462	0.8355	0.8247	0.8140	0.8034	0.7929	0.7825	0.7722	0.7621	0.7522	0.7424	0.7329	0.7236	0.7144	0.7055
5	0.8266	0.8131	0.7995	0.7860	0.7725	0.7591	0.7459	0.7330	0.7202	0.7078	0.6956	0.6837	0.6722	0.6609	0.6499
6	0.8100	0.7937	0.7773	0.7609	0.7446	0.7285	0.7126	0.6970	0.6818	0.6670	0.6525	0.6385	0.6249	0.6118	0.5991
7	0.7956	0.7767	0.7575	0.7382	0.7191	0.7003	0.6818	0.6637	0.6461	0.6291	0.6126	0.5967	0.5814	0.5666	0.5524
8	0.7830	0.7613	0.7394	0.7174	0.6955	0.6740	0.6530	0.6326	0.6128	0.5937	0.5754	0.5578	0.5410	0.5249	0.5095
9	0.7717	0.7474	0.7227	0.6980	0.6734	0.6493	0.6259	0.6033	0.5815	0.5606	0.5406	0.5216	0.5035	0.4863	0.4699
10	0.7616	0.7347	0.7072	0.6798	0.6526	0.6260	0.6003	0.5756	0.5520	0.5295	0.5081	0.4879	0.4687	0.4506	0.4335

P=0.50
K=0.90

YEAR*	0	5	10	15	20	25	30	35	40	45	50	55	60	65	70
0	1.0000	1.0000	1.0000	1.0000	1.0000	1.0000	1.0000	1.0000	1.0000	1.0000	1.0000	1.0000	1.0000	1.0000	1.0000
1	0.9704	0.9693	0.9681	0.9669	0.9658	0.9647	0.9636	0.9625	0.9615	0.9605	0.9594	0.9584	0.9574	0.9565	0.9555
2	0.9500	0.9474	0.9448	0.9422	0.9396	0.9371	0.9346	0.9321	0.9296	0.9272	0.9248	0.9225	0.9201	0.9178	0.9155
3	0.9344	0.9303	0.9261	0.9220	0.9179	0.9138	0.9097	0.9056	0.9016	0.8977	0.8938	0.8899	0.8861	0.8823	0.8786
4	0.9219	0.9162	0.9105	0.9047	0.8989	0.8932	0.8874	0.8818	0.8761	0.8705	0.8650	0.8596	0.8542	0.8490	0.8438
5	0.9115	0.9042	0.8968	0.8894	0.8819	0.8744	0.8670	0.8596	0.8523	0.8451	0.8380	0.8310	0.8241	0.8174	0.8108
6	0.9025	0.8936	0.8845	0.8755	0.8663	0.8571	0.8479	0.8389	0.8299	0.8210	0.8123	0.8038	0.7954	0.7873	0.7792
7	0.8947	0.8842	0.8735	0.8627	0.8517	0.8408	0.8299	0.8191	0.8085	0.7980	0.7877	0.7777	0.7679	0.7584	0.7491
8	0.8877	0.8757	0.8633	0.8507	0.8380	0.8253	0.8126	0.8001	0.7879	0.7758	0.7641	0.7526	0.7415	0.7307	0.7201
9	0.8815	0.8679	0.8538	0.8394	0.8249	0.8104	0.7960	0.7819	0.7680	0.7545	0.7413	0.7284	0.7160	0.7040	0.6924
10	0.8758	0.8606	0.8448	0.8287	0.8124	0.7961	0.7800	0.7642	0.7488	0.7338	0.7192	0.7051	0.6915	0.6783	0.6657

P=0.50
K=0.95

* GROWTH

YEAR*	0	5	10	15	20	25	30	35	40	45	50	55	60	65	70
0	1.0000	1.0000	1.0000	1.0000	1.0000	1.0000	1.0000	1.0000	1.0000	1.0000	1.0000	1.0000	1.0000	1.0000	1.0000
1	0.7981	0.7909	0.7839	0.7771	0.7704	0.7640	0.7576	0.7514	0.7454	0.7395	0.7338	0.7281	0.7226	0.7173	0.7120
2	0.6826	0.6690	0.6558	0.6429	0.6303	0.6181	0.6063	0.5943	0.5836	0.5727	0.5622	0.5519	0.5420	0.5323	0.5230
3	0.6056	0.5865	0.5679	0.5499	0.5325	0.5157	0.4995	0.4839	0.4689	0.4544	0.4405	0.4272	0.4144	0.4020	0.3902
4	0.5496	0.5257	0.5026	0.4803	0.4589	0.4385	0.4189	0.4003	0.3826	0.3657	0.3497	0.3346	0.3202	0.3065	0.2936
5	0.5065	0.4784	0.4513	0.4254	0.4008	0.3774	0.3554	0.3347	0.3152	0.2969	0.2799	0.2539	0.2490	0.2350	0.2220
6	0.4721	0.4401	0.4095	0.3806	0.3533	0.3277	0.3039	0.2818	0.2613	0.2425	0.2251	0.2090	0.1943	0.1808	0.1683
7	0.4437	0.4083	0.3745	0.3430	0.3136	0.2863	0.2613	0.2385	0.2176	0.1987	0.1816	0.1560	0.1520	0.1393	0.1278
8	0.4199	0.3812	0.3443	0.3109	0.2797	0.2513	0.2256	0.2025	0.1318	0.1633	0.1468	0.1321	0.1191	0.1075	0.0972
9	0.3994	0.3578	0.3189	0.2831	0.2506	0.2213	0.1953	0.1723	0.1521	0.1344	0.1188	0.1052	0.0934	0.0830	0.0739
10	0.3317	0.3373	0.2952	0.2587	0.2251	0.1955	0.1695	0.1470	0.1275	0.1107	0.0963	0.0839	0.0732	0.0641	0.0562

P=0.55
K=0.70

YEAR	0	5	10	15	20	25	30	35	40	45	50	55	60	65	70
0	1.0000	1.0000	1.0000	1.0000	1.0000	1.0000	1.0000	1.0000	1.0000	1.0000	1.0000	1.0000	1.0000	1.0000	1.0000
1	0.8337	0.8276	0.8217	0.8159	0.8103	0.8048	0.7994	0.7942	0.7890	0.7840	0.7791	0.7742	0.7695	0.7649	0.7604
2	0.7350	0.7231	0.7115	0.7002	0.6892	0.6784	0.6679	0.6577	0.6477	0.6379	0.6284	0.6192	0.6102	0.6014	0.5928
3	0.6673	0.6503	0.6336	0.6174	0.6016	0.5862	0.5713	0.5569	0.5429	0.5293	0.5162	0.5036	0.4913	0.4795	0.4681
4	0.6171	0.5953	0.5741	0.5535	0.5336	0.5143	0.4957	0.4779	0.4607	0.4443	0.4285	0.4135	0.3991	0.3853	0.3722
5	0.5778	0.5517	0.5264	0.5019	0.4783	0.4557	0.4342	0.4136	0.3941	0.3756	0.3580	0.3414	0.3258	0.3110	0.2970
6	0.5459	0.5158	0.4867	0.4588	0.4320	0.4067	0.3827	0.3600	0.3388	0.3189	0.3003	0.2830	0.2668	0.2517	0.2376
7	0.5193	0.4855	0.4530	0.4219	0.3924	0.3647	0.3388	0.3147	0.2923	0.2716	0.2526	0.2350	0.2188	0.2040	0.1903
8	0.4966	0.4594	0.4237	0.3897	0.3579	0.3283	0.3009	0.2758	0.2528	0.2318	0.2127	0.1954	0.1797	0.1655	0.1525
9	0.4770	0.4365	0.3978	0.3614	0.3275	0.2963	0.2679	0.2422	0.2190	0.1981	0.1794	0.1527	0.1477	0.1343	0.1223
10	0.4598	0.4162	0.3748	0.3360	0.3004	0.2680	0.2389	0.2130	0.1899	0.1694	0.1514	0.1355	0.1214	0.1090	0.0981

P=0.55
K=0.75

*
 *
  * GROWTH
   *

YEAR*	0	5	10	15	20	25	30	35	40	45	50	55	60	65	70
0	1.0000	1.0000	1.0000	1.0000	1.0000	1.0000	1.0000	1.0000	1.0000	1.0000	1.0000	1.0000	1.0000	1.0000	1.0000
1	0.8684	0.8635	0.8587	0.8540	0.8495	0.8450	0.8406	0.8363	0.8321	0.8280	0.8239	0.8200	0.8161	0.8123	0.8086
2	0.7875	0.7777	0.7680	0.7585	0.7492	0.7401	0.7312	0.7225	0.7140	0.7056	0.6975	0.6895	0.6817	0.6741	0.6665
3	0.7307	0.7162	0.7019	0.6879	0.6742	0.6608	0.6478	0.6350	0.6226	0.6105	0.5988	0.5874	0.5763	0.5655	0.5550
4	0.6877	0.6688	0.6502	0.6321	0.6143	0.5970	0.5802	0.5640	0.5482	0.5330	0.5183	0.5041	0.4904	0.4772	0.4646
5	0.6534	0.6305	0.6079	0.5858	0.5644	0.5436	0.5235	0.5042	0.4856	0.4678	0.4508	0.4345	0.4190	0.4042	0.3900
6	0.6253	0.5934	0.5721	0.5464	0.5215	0.4976	0.4747	0.4528	0.4319	0.4121	0.3933	0.3756	0.3588	0.3430	0.3280
7	0.6015	0.5710	0.5410	0.5120	0.4840	0.4573	0.4319	0.4078	0.3852	0.3639	0.3439	0.3252	0.3077	0.2914	0.2761
8	0.5811	0.5470	0.5137	0.4815	0.4507	0.4215	0.3940	0.3682	0.3441	0.3218	0.3011	0.2819	0.2641	0.2477	0.2326
9	0.5632	0.5258	0.4892	0.4541	0.4207	0.3893	0.3600	0.3329	0.3079	0.2849	0.2638	0.2445	0.2268	0.2107	0.1959
10	0.5474	0.5067	0.4671	0.4292	0.3934	0.3601	0.3294	0.3013	0.2756	0.2523	0.2312	0.2121	0.1949	0.1793	0.1651

****
ρ=0.55
****
K=0.30
****

YEAR*	0	5	10	15	20	25	30	35	40	45	50	55	60	65	70
0	1.0000	1.0000	1.0000	1.0000	1.0000	1.0000	1.0000	1.0000	1.0000	1.0000	1.0000	1.0000	1.0000	1.0000	1.0000
1	0.9023	0.8986	0.8950	0.8914	0.8880	0.8845	0.8812	0.8779	0.8747	0.8715	0.8684	0.8654	0.8624	0.8595	0.8566
2	0.8403	0.8327	0.8251	0.8177	0.8103	0.8032	0.7961	0.7892	0.7824	0.7757	0.7692	0.7628	0.7565	0.7503	0.7443
3	0.7957	0.7842	0.7728	0.7615	0.7504	0.7396	0.7289	0.7184	0.7082	0.6981	0.6883	0.6787	0.6694	0.6602	0.6513
4	0.7613	0.7460	0.7309	0.7160	0.7013	0.6868	0.6727	0.6589	0.6455	0.6324	0.6196	0.6072	0.5952	0.5835	0.5721
5	0.7335	0.7146	0.6959	0.6774	0.6593	0.6415	0.6242	0.6073	0.5909	0.5751	0.5598	0.5450	0.5307	0.5169	0.5037
6	0.7104	0.6880	0.6658	0.6439	0.6224	0.6015	0.5812	0.5615	0.5426	0.5243	0.5068	0.4901	0.4740	0.4587	0.4440
7	0.6906	0.6649	0.6393	0.6141	0.5895	0.5656	0.5425	0.5204	0.4991	0.4789	0.4596	0.4413	0.4238	0.4073	0.3917
8	0.6734	0.6444	0.6156	0.5873	0.5596	0.5330	0.5074	0.4830	0.4598	0.4379	0.4171	0.3976	0.3792	0.3619	0.3456
9	0.6583	0.6261	0.5941	0.5627	0.5323	0.5030	0.4752	0.4488	0.4240	0.4007	0.3789	0.3585	0.3394	0.3217	0.3051
10	0.6448	0.6095	0.5744	0.5400	0.5069	0.4753	0.4454	0.4174	0.3912	0.3668	0.3442	0.3233	0.3039	0.2859	0.2694

****
ρ=0.55
****
K=0.85
****

*

*

* GROWTH

*

YEAR*	0	5	10	15	20	25	30	35	40	45	50	55	60	65	70
0	1.0000	1.0000	1.0000	1.0000	1.0000	1.0000	1.0000	1.0000	1.0000	1.0000	1.0000	1.0000	1.0000	1.0000	1.0000
1	0.9356	0.9331	0.9306	0.9282	0.9259	0.9235	0.9213	0.9191	0.9169	0.9147	0.9126	0.9105	0.9085	0.9065	0.9045
2	0.8934	0.8880	0.8828	0.8776	0.8726	0.8675	0.8626	0.8577	0.8529	0.8482	0.8436	0.8390	0.8345	0.8301	0.8257
3	0.8623	0.8542	0.8461	0.8381	0.8302	0.8223	0.8146	0.8070	0.7995	0.7922	0.7849	0.7778	0.7709	0.7640	0.7573
4	0.8379	0.8270	0.8161	0.8052	0.7945	0.7839	0.7734	0.7630	0.7529	0.7430	0.7332	0.7237	0.7143	0.7052	0.6963
5	0.8180	0.8043	0.7906	0.7769	0.7633	0.7499	0.7367	0.7237	0.7110	0.6986	0.6865	0.6747	0.6632	0.6520	0.6411
6	0.8011	0.7847	0.7682	0.7517	0.7354	0.7193	0.7034	0.6879	0.6727	0.6580	0.6437	0.6298	0.6163	0.6033	0.5908
7	0.7866	0.7675	0.7482	0.7290	0.7099	0.6911	0.6727	0.6543	0.6373	0.6204	0.6041	0.5884	0.5732	0.5586	0.5446
8	0.7739	0.7521	0.7301	0.7082	0.6864	0.6650	0.6441	0.6239	0.6043	0.5854	0.5673	0.5500	0.5333	0.5174	0.5022
9	0.7626	0.7382	0.7135	0.6888	0.6644	0.6405	0.6173	0.5949	0.5733	0.5527	0.5330	0.5142	0.4963	0.4794	0.4632
10	0.7524	0.7254	0.6981	0.6707	0.6437	0.6174	0.5920	0.5675	0.5442	0.5220	0.5009	0.4809	0.4620	0.4441	0.4273

ρ=0.55

K=0.90

YEAR*	0	5	10	15	20	25	30	35	40	45	50	55	60	65	70
0	1.0000	1.0000	1.0000	1.0000	1.0000	1.0000	1.0000	1.0000	1.0000	1.0000	1.0000	1.0000	1.0000	1.0000	1.0000
1	0.9681	0.9658	0.9656	0.9644	0.9632	0.9620	0.9609	0.9597	0.9586	0.9575	0.9565	0.9554	0.9544	0.9533	0.9523
2	0.9466	0.9433	0.9411	0.9384	0.9358	0.9332	0.9306	0.9280	0.9255	0.9230	0.9205	0.9181	0.9157	0.9133	0.9110
3	0.9304	0.9261	0.9219	0.9176	0.9134	0.9092	0.9050	0.9009	0.8968	0.8928	0.8888	0.8849	0.8810	0.8772	0.8734
4	0.9175	0.9117	0.9058	0.8999	0.8940	0.8882	0.8824	0.8766	0.8709	0.8653	0.8598	0.8543	0.8489	0.8436	0.8384
5	0.9068	0.8994	0.8919	0.8843	0.8768	0.8693	0.8618	0.8544	0.8470	0.8398	0.8327	0.8256	0.8188	0.8120	0.8054
6	0.8977	0.8887	0.8795	0.8703	0.8610	0.8518	0.8426	0.8335	0.8245	0.8157	0.8070	0.7984	0.7901	0.7819	0.7740
7	0.8897	0.8791	0.8683	0.8574	0.8464	0.8354	0.8245	0.8137	0.8031	0.7926	0.7824	0.7724	0.7627	0.7532	0.7439
8	0.8827	0.8705	0.8580	0.8454	0.8326	0.8199	0.8072	0.7948	0.7825	0.7706	0.7589	0.7475	0.7364	0.7256	0.7151
9	0.8764	0.8626	0.8484	0.8340	0.8195	0.8050	0.7907	0.7766	0.7628	0.7493	0.7361	0.7234	0.7110	0.6991	0.6875
10	0.8707	0.8553	0.8395	0.8233	0.8070	0.7908	0.7747	0.7590	0.7436	0.7287	0.7142	0.7002	0.6866	0.6736	0.6610

ρ=0.55

K=0.95

YEAR* GROWTH	0	5	10	15	20	25	30	35	40	45	50	55	60	65	70
0	1.0000	1.0000	1.0000	1.0000	1.0000	1.0000	1.0000	1.0000	1.0000	1.0000	1.0000	1.0000	1.0000	1.0000	1.0000
1	0.7852	0.7777	0.7704	0.7634	0.7565	0.7498	0.7433	0.7369	0.7307	0.7246	0.7187	0.7130	0.7073	0.7018	0.6964
2	0.6665	0.6527	0.6392	0.6262	0.6135	0.6013	0.5894	0.5778	0.5666	0.5558	0.5453	0.5351	0.5252	0.5156	0.5063
3	0.5887	0.5695	0.5510	0.5331	0.5158	0.4991	0.4831	0.4677	0.4528	0.4386	0.4250	0.4119	0.3993	0.3873	0.3757
4	0.5327	0.5090	0.4861	0.4641	0.4430	0.4229	0.4038	0.3856	0.3683	0.3519	0.3363	0.3216	0.3076	0.2944	0.2819
5	0.4900	0.4622	0.4355	0.4101	0.3860	0.3633	0.3418	0.3217	0.3028	0.2851	0.2686	0.2531	0.2387	0.2253	0.2128
6	0.4560	0.4246	0.3946	0.3663	0.3397	0.3149	0.2918	0.2705	0.2507	0.2325	0.2157	0.2003	0.1861	0.1731	0.1612
7	0.4281	0.3934	0.3605	0.3298	0.3012	0.2748	0.2507	0.2286	0.2086	0.1904	0.1739	0.1590	0.1455	0.1333	0.1223
8	0.4047	0.3670	0.3315	0.2986	0.2685	0.2410	0.2162	0.1940	0.1741	0.1563	0.1405	0.1264	0.1139	0.1028	0.0929
9	0.3847	0.3442	0.3064	0.2717	0.2403	0.2121	0.1871	0.1650	0.1456	0.1286	0.1137	0.1007	0.0893	0.0794	0.0707
10	0.3674	0.3242	0.2843	0.2481	0.2158	0.1872	0.1623	0.1407	0.1220	0.1059	0.0921	0.0802	0.0700	0.0613	0.0538

ρ=0.60
K=0.70

YEAR	0	5	10	15	20	25	30	35	40	45	50	55	60	65	70
0	1.0000	1.0000	1.0000	1.0000	1.0000	1.0000	1.0000	1.0000	1.0000	1.0000	1.0000	1.0000	1.0000	1.0000	1.0000
1	0.8228	0.8165	0.8103	0.8043	0.7984	0.7927	0.7872	0.7817	0.7764	0.7712	0.7661	0.7612	0.7563	0.7516	0.7469
2	0.7209	0.7088	0.6970	0.6855	0.6743	0.6634	0.6528	0.6425	0.6324	0.6227	0.5131	0.6039	0.5948	0.5861	0.5775
3	0.6522	0.6351	0.6183	0.6020	0.5862	0.5709	0.5561	0.5417	0.5278	0.5144	0.5015	0.4390	0.4769	0.4653	0.4540
4	0.6018	0.5800	0.5589	0.5384	0.5186	0.4995	0.4812	0.4636	0.4468	0.4306	0.4152	0.4005	0.3864	0.3730	0.3601
5	0.5625	0.5366	0.5115	0.4873	0.4641	0.4419	0.4207	0.4006	0.3815	0.3634	0.3463	0.3302	0.3150	0.3006	0.2870
6	0.5308	0.5011	0.4724	0.4449	0.4186	0.3938	0.3703	0.3483	0.3276	0.3083	0.2902	0.2734	0.2577	0.2430	0.2294
7	0.5045	0.4712	0.4392	0.4087	0.3799	0.3528	0.3276	0.3041	0.2824	0.2624	0.2439	0.2269	0.2113	0.1969	0.1837
8	0.4821	0.4455	0.4104	0.3773	0.3462	0.3174	0.2908	0.2664	0.2441	0.2238	0.2054	0.1886	0.1734	0.1597	0.1472
9	0.4628	0.4230	0.3852	0.3496	0.3166	0.2863	0.2588	0.2338	0.2114	0.1912	0.1731	0.1570	0.1425	0.1296	0.1180
10	0.4459	0.4032	0.3626	0.3249	0.2903	0.2589	0.2307	0.2056	0.1833	0.1635	0.1461	0.1307	0.1171	0.1052	0.0946

ρ=0.60
K=0.75

*
*
* GROWTH
*

YEAR*	0	5	10	15	20	25	30	35	40	45	50	55	60	65	70
0	1.0000	1.0000	1.0000	1.0000	1.0000	1.0000	1.0000	1.0000	1.0000	1.0000	1.0000	1.0000	1.0000	1.0000	1.0000
1	0.8596	0.8545	0.8495	0.8446	0.8398	0.8351	0.8306	0.8261	0.8218	0.8175	0.8133	0.8092	0.8052	0.8013	0.7974
2	0.7758	0.7657	0.7558	0.7461	0.7367	0.7274	0.7184	0.7095	0.7009	0.6925	0.6842	0.6762	0.6684	0.6607	0.6532
3	0.7179	0.7032	0.6887	0.6746	0.6609	0.6474	0.6343	0.6216	0.6092	0.5972	0.5855	0.5741	0.5631	0.5524	0.5420
4	0.6744	0.6554	0.6368	0.6186	0.6009	0.5837	0.5670	0.5509	0.5353	0.5202	0.5057	0.4917	0.4783	0.4653	0.4529
5	0.6400	0.6170	0.5945	0.5726	0.5513	0.5307	0.5109	0.4913	0.4736	0.4561	0.4393	0.4234	0.4082	0.3936	0.3798
6	0.6118	0.5851	0.5589	0.5335	0.5090	0.4854	0.4628	0.4413	0.4208	0.4014	0.3830	0.3657	0.3493	0.3338	0.3192
7	0.5882	0.5578	0.5282	0.4995	0.4720	0.4457	0.4208	0.3972	0.3751	0.3542	0.3347	0.3165	0.2994	0.2835	0.2686
8	0.5678	0.5341	0.5012	0.4695	0.4392	0.4106	0.3836	0.3584	0.3350	0.3131	0.2929	0.2742	0.2569	0.2410	0.2262
9	0.5501	0.5131	0.4771	0.4425	0.4098	0.3791	0.3505	0.3240	0.2996	0.2771	0.2566	0.2378	0.2206	0.2049	0.1906
10	0.5345	0.4943	0.4553	0.4181	0.3831	0.3506	0.3206	0.2932	0.2682	0.2455	0.2249	0.2063	0.1895	0.1743	0.1606

ρ=0.60
K=0.80

YEAR*	0	5	10	15	20	25	30	35	40	45	50	55	60	65	70
0	1.0000	1.0000	1.0000	1.0000	1.0000	1.0000	1.0000	1.0000	1.0000	1.0000	1.0000	1.0000	1.0000	1.0000	1.0000
1	0.8957	0.8918	0.8880	0.8842	0.8806	0.8770	0.8735	0.8701	0.8668	0.8635	0.8603	0.8571	0.8540	0.8510	0.8480
2	0.8312	0.8233	0.8155	0.8079	0.8004	0.7931	0.7859	0.7789	0.7720	0.7652	0.7585	0.7520	0.7457	0.7394	0.7333
3	0.7855	0.7738	0.7622	0.7508	0.7396	0.7286	0.7178	0.7073	0.6970	0.6869	0.6771	0.6675	0.6582	0.6491	0.6402
4	0.7506	0.7351	0.7199	0.7048	0.6901	0.6756	0.6615	0.6478	0.6344	0.6213	0.6086	0.5963	0.5844	0.5728	0.5616
5	0.7225	0.7035	0.6847	0.6662	0.6481	0.6304	0.6132	0.5964	0.5802	0.5645	0.5494	0.5347	0.5207	0.5071	0.4941
6	0.6992	0.6768	0.6546	0.6328	0.6115	0.5907	0.5706	0.5511	0.5324	0.5144	0.4971	0.4806	0.4648	0.4497	0.4353
7	0.6794	0.6537	0.6282	0.6032	0.5788	0.5552	0.5324	0.5105	0.4896	0.4696	0.4506	0.4326	0.4155	0.3993	0.3839
8	0.6622	0.6333	0.6047	0.5766	0.5492	0.5229	0.4977	0.4737	0.4509	0.4293	0.4089	0.3897	0.3717	0.3547	0.3387
9	0.6471	0.6151	0.5833	0.5523	0.5222	0.4934	0.4660	0.4400	0.4156	0.3927	0.3713	0.3513	0.3326	0.3152	0.2990
10	0.6337	0.5986	0.5638	0.5299	0.4972	0.4661	0.4367	0.4092	0.3834	0.3595	0.3373	0.3168	0.2978	0.2802	0.2639

ρ=0.60
K=0.85

GROWTH

P=0.60, K=0.90

YEAR	0	5	10	15	20	25	30	35	40	45	50	55	60	65	70
0	1.0000	1.0000	1.0000	1.0000	1.0000	1.0000	1.0000	1.0000	1.0000	1.0000	1.0000	1.0000	1.0000	1.0000	1.0000
1	0.9311	0.9234	0.9259	0.9233	0.9209	0.9185	0.9161	0.9138	0.9115	0.9092	0.9070	0.9049	0.9028	0.9007	0.8986
2	0.3371	0.8816	0.8762	0.8709	0.8656	0.8605	0.8554	0.8504	0.8455	0.8407	0.8360	0.8313	0.8268	0.8223	0.8179
3	0.8551	0.8468	0.8386	0.8304	0.8224	0.8144	0.8066	0.7989	0.7914	0.7839	0.7766	0.7695	0.7625	0.7556	0.7489
4	0.8303	0.8191	0.8081	0.7971	0.7863	0.7755	0.7650	0.7546	0.7445	0.7345	0.7248	0.7152	0.7059	0.6968	0.6880
5	0.3100	0.7961	0.7823	0.7685	0.7549	0.7415	0.7283	0.7153	0.7026	0.6903	0.6782	0.5564	0.6550	0.6439	0.6331
6	0.7930	0.7764	0.7598	0.7433	0.7270	0.7108	0.6950	0.6796	0.6645	0.6499	0.6357	0.6219	0.6086	0.5957	0.5832
7	0.7783	0.7591	0.7398	0.7206	0.7015	0.6828	0.6645	0.6467	0.6294	0.6126	0.5964	0.5809	0.5659	0.5514	0.5376
8	0.7655	0.7437	0.7217	0.6998	0.6781	0.6568	0.6361	0.6160	0.5966	0.5780	0.5600	0.5429	0.5264	0.5107	0.4957
9	0.7542	0.7297	0.7051	0.6805	0.6563	0.6325	0.6095	0.5873	0.5660	0.5456	0.5261	0.5075	0.4899	0.4731	0.4572
10	0.7439	0.7170	0.6897	0.6625	0.6357	0.6096	0.5844	0.5603	0.5372	0.5152	0.4944	0.4746	0.4560	0.4383	0.4217

P=0.60, K=0.95

YEAR	0	5	10	15	20	25	30	35	40	45	50	55	60	65	70
0	1.0000	1.0000	1.0000	1.0000	1.0000	1.0000	1.0000	1.0000	1.0000	1.0000	1.0000	1.0000	1.0000	1.0000	1.0000
1	0.9658	0.9645	0.9632	0.9619	0.9607	0.9594	0.9582	0.9570	0.9559	0.9547	0.9536	0.9525	0.9514	0.9504	0.9493
2	0.9433	0.9405	0.9377	0.9349	0.9322	0.9295	0.9268	0.9242	0.9216	0.9190	0.9165	0.9140	0.9115	0.9091	0.9067
3	0.9266	0.9222	0.9179	0.9135	0.9092	0.9049	0.9007	0.8965	0.8923	0.3882	0.8842	0.3802	0.8763	0.8725	0.8687
4	0.9134	0.9074	0.9015	0.8955	0.8895	0.8836	0.8777	0.8719	0.8662	0.3605	0.8549	0.8495	0.8441	0.8387	0.8335
5	0.9025	0.3949	0.8873	0.8797	0.8721	0.8645	0.8570	0.3495	0.8421	0.8349	0.8277	0.8207	0.8138	0.8071	0.8005
6	0.3932	0.8841	0.8748	0.8655	0.8562	0.8469	0.8377	0.8286	0.8196	0.8107	0.8021	0.7935	0.7852	0.7771	0.7691
7	0.3851	0.8744	0.8635	0.8525	0.8415	0.8305	0.8196	0.8088	0.7982	0.7878	0.7776	0.7576	0.7579	0.7484	0.7392
8	0.8780	0.8657	0.8532	0.8405	0.8277	0.8150	0.8023	0.7899	0.7777	0.7657	0.7541	0.7427	0.7317	0.7210	0.7106
9	0.3717	0.8578	0.8436	0.8291	0.8146	0.8001	0.7858	0.7718	0.7580	0.7445	0.7315	0.7188	0.7065	0.6946	0.6831
10	0.8659	0.8505	0.8346	0.8184	0.8021	0.7859	0.7699	0.7542	0.7389	0.7241	0.7097	0.6957	0.6823	0.6693	0.6568

GROWTH

YEAR	0	5	10	15	20	25	30	35	40	45	50	55	60	65	70
0	1.0000	1.0000	1.0000	1.0000	1.0000	1.0000	1.0000	1.0000	1.0000	1.0000	1.0000	1.0000	1.0000	1.0000	1.0000
1	0.7728	0.7651	0.7576	0.7503	0.7433	0.7364	0.7297	0.7231	0.7168	0.7106	0.7045	0.6987	0.6929	0.6873	0.6818
2	0.6514	0.6374	0.6239	0.6107	0.5980	0.5857	0.5738	0.5622	0.5510	0.5402	0.5297	0.5196	0.5098	0.5003	0.4910
3	0.5731	0.5540	0.5354	0.5176	0.5005	0.4840	0.4681	0.4529	0.4383	0.4243	0.4109	0.3981	0.3858	0.3740	0.3627
4	0.5173	0.4937	0.4711	0.4494	0.4287	0.4089	0.3902	0.3723	0.3554	0.3394	0.3243	0.3099	0.2964	0.2836	0.2714
5	0.4750	0.4475	0.4213	0.3964	0.3728	0.3506	0.3297	0.3100	0.2917	0.2745	0.2585	0.2436	0.2297	0.2167	0.2046
6	0.4414	0.4105	0.3812	0.3535	0.3276	0.3035	0.2811	0.2604	0.2412	0.2236	0.2074	0.1925	0.1789	0.1664	0.1548
7	0.4140	0.3800	0.3479	0.3179	0.2901	0.2646	0.2412	0.2199	0.2005	0.1830	0.1671	0.1527	0.1398	0.1280	0.1175
8	0.3911	0.3542	0.3196	0.2877	0.2584	0.2319	0.2079	0.1865	0.1673	0.1502	0.1349	0.1214	0.1094	0.0987	0.0892
9	0.3715	0.3320	0.2952	0.2616	0.2312	0.2040	0.1799	0.1586	0.1399	0.1235	0.1092	0.0967	0.0857	0.0762	0.0678
10	0.3546	0.3125	0.2738	0.2337	0.2075	0.1800	0.1559	0.1351	0.1172	0.1017	0.0884	0.0770	0.0672	0.0588	0.0516

$\rho=0.65$

$K=0.70$

YEAR	0	5	10	15	20	25	30	35	40	45	50	55	60	65	70
0	1.0000	1.0000	1.0000	1.0000	1.0000	1.0000	1.0000	1.0000	1.0000	1.0000	1.0000	1.0000	1.0000	1.0000	1.0000
1	0.8123	0.8058	0.7994	0.7932	0.7872	0.7813	0.7755	0.7699	0.7645	0.7591	0.7539	0.7488	0.7439	0.7390	0.7342
2	0.7077	0.6954	0.6835	0.6719	0.6605	0.6495	0.6389	0.6285	0.6184	0.6086	0.5990	0.5897	0.5807	0.5720	0.5635
3	0.6383	0.6210	0.6042	0.5879	0.5722	0.5569	0.5421	0.5279	0.5141	0.5009	0.4881	0.4757	0.4638	0.4524	0.4413
4	0.5876	0.5659	0.5449	0.5246	0.5050	0.4862	0.4681	0.4508	0.4342	0.4183	0.4032	0.3888	0.3750	0.3618	0.3493
5	0.5485	0.5228	0.4980	0.4741	0.4512	0.4294	0.4086	0.3889	0.3702	0.3525	0.3359	0.3201	0.3053	0.2913	0.2781
6	0.5171	0.4877	0.4594	0.4323	0.4066	0.3822	0.3593	0.3378	0.3176	0.2988	0.2812	0.2648	0.2495	0.2353	0.2221
7	0.4910	0.4582	0.4267	0.3968	0.3686	0.3422	0.3176	0.2947	0.2736	0.2541	0.2362	0.2197	0.2045	0.1906	0.1778
8	0.4690	0.4329	0.3985	0.3661	0.3357	0.3076	0.2818	0.2580	0.2364	0.2167	0.1988	0.1826	0.1678	0.1545	0.1424
9	0.4499	0.4109	0.3738	0.3390	0.3069	0.2774	0.2506	0.2264	0.2046	0.1851	0.1676	0.1519	0.1379	0.1254	0.1141
10	0.4333	0.3914	0.3518	0.3150	0.2813	0.2508	0.2234	0.1990	0.1774	0.1582	0.1414	0.1265	0.1133	0.1018	0.0915

$\rho=0.65$

$K=0.75$

*
*
* GROWTH
*

YEAR*	0	5	10	15	20	25	30	35	40	45	50	55	60	65	70
0	1.0000	1.0000	1.0000	1.0000	1.0000	1.0000	1.0000	1.0000	1.0000	1.0000	1.0000	1.0000	1.0000	1.0000	1.0000
1	0.8511	0.8458	0.8406	0.8355	0.8306	0.8258	0.8210	0.8164	0.8119	0.8075	0.8032	0.7990	0.7949	0.7909	0.7869
2	0.7648	0.7545	0.7444	0.7346	0.7249	0.7156	0.7064	0.6975	0.6888	0.6803	0.6720	0.6639	0.6560	0.6484	0.6409
3	0.7059	0.6911	0.6765	0.6623	0.6485	0.6350	0.6220	0.6092	0.5969	0.5849	0.5733	0.5620	0.5511	0.5405	0.5302
4	0.6621	0.6430	0.6244	0.6063	0.5886	0.5715	0.5550	0.5390	0.5235	0.5087	0.4943	0.4805	0.4673	0.4545	0.4423
5	0.6276	0.6047	0.5823	0.5605	0.5394	0.5190	0.4994	0.4806	0.4626	0.4454	0.4290	0.4133	0.3984	0.3841	0.3706
6	0.5995	0.5729	0.5469	0.5218	0.4975	0.4743	0.4520	0.4309	0.4108	0.3918	0.3738	0.3568	0.3407	0.3256	0.3113
7	0.5760	0.5459	0.5165	0.4882	0.4611	0.4353	0.4108	0.3877	0.3659	0.3456	0.3265	0.3086	0.2920	0.2764	0.2619
8	0.5558	0.5224	0.4899	0.4586	0.4289	0.4008	0.3744	0.3497	0.3267	0.3054	0.2856	0.2574	0.2505	0.2349	0.2205
9	0.5332	0.5016	0.4661	0.4321	0.4000	0.3699	0.3419	0.3160	0.2921	0.2702	0.2502	0.2318	0.2151	0.1997	0.1857
10	0.5227	0.4831	0.4447	0.4032	0.3739	0.3420	0.3127	0.2859	0.2614	0.2393	0.2192	0.2011	0.1847	0.1699	0.1565

P=0.65
K=0.80

YEAR	0	5	10	15	20	25	30	35	40	45	50	55	60	65	70
0	1.0000	1.0000	1.0000	1.0000	1.0000	1.0000	1.0000	1.0000	1.0000	1.0000	1.0000	1.0000	1.0000	1.0000	1.0000
1	0.8892	0.8852	0.8812	0.8773	0.8735	0.8698	0.8662	0.8627	0.8592	0.8558	0.8525	0.8493	0.8461	0.8429	0.8399
2	0.8226	0.8145	0.8066	0.7988	0.7911	0.7837	0.7764	0.7692	0.7622	0.7553	0.7486	0.7421	0.7356	0.7294	0.7232
3	0.7760	0.7640	0.7523	0.7408	0.7295	0.7184	0.7076	0.6970	0.6867	0.6766	0.6668	0.6572	0.6479	0.6388	0.6300
4	0.7406	0.7250	0.7096	0.6946	0.6798	0.6653	0.6513	0.6375	0.6242	0.6112	0.5986	0.5864	0.5746	0.5631	0.5520
5	0.7123	0.6932	0.6744	0.6560	0.6379	0.6203	0.6031	0.5865	0.5704	0.5549	0.5399	0.5255	0.5116	0.4982	0.4853
6	0.6889	0.6665	0.6444	0.6226	0.6014	0.5808	0.5609	0.5416	0.5231	0.5054	0.4883	0.4721	0.4565	0.4416	0.4274
7	0.6691	0.6434	0.6181	0.5932	0.5690	0.5456	0.5231	0.5015	0.4809	0.4612	0.4425	0.4248	0.4079	0.3920	0.3769
8	0.6519	0.6232	0.5947	0.5668	0.5398	0.5138	0.4889	0.4652	0.4427	0.4215	0.4015	0.3826	0.3649	0.3482	0.3325
9	0.6369	0.6050	0.5735	0.5428	0.5131	0.4846	0.4576	0.4321	0.4081	0.3856	0.3645	0.3448	0.3265	0.3094	0.2935
10	0.6235	0.5887	0.5542	0.5207	0.4884	0.4577	0.4288	0.4017	0.3764	0.3529	0.3311	0.3109	0.2923	0.2750	0.2590

P=0.65
K=0.85

GROWTH

YEAR	0	5	10	15	20	25	30	35	40	45	50	55	60	65	70
0	1.0000	1.0000	1.0000	1.0000	1.0000	1.0000	1.0000	1.0000	1.0000	1.0000	1.0000	1.0000	1.0000	1.0000	1.0000
1	0.9267	0.9240	0.9213	0.9187	0.9161	0.9136	0.9111	0.9087	0.9063	0.9040	0.9017	0.8995	0.8973	0.8951	0.8930
2	0.8811	0.8754	0.8699	0.8645	0.8591	0.8538	0.8487	0.8436	0.8386	0.8337	0.8289	0.8242	0.8195	0.8150	0.8105
3	0.8484	0.8399	0.8315	0.8232	0.8151	0.8070	0.7991	0.7914	0.7838	0.7763	0.7690	0.7618	0.7548	0.7479	0.7411
4	0.8231	0.8118	0.8006	0.7896	0.7786	0.7679	0.7573	0.7469	0.7367	0.7268	0.7170	0.7075	0.6982	0.6892	0.6803
5	0.8026	0.7886	0.7746	0.7608	0.7472	0.7337	0.7205	0.7076	0.6949	0.6826	0.6706	0.6589	0.6476	0.6365	0.6258
6	0.7854	0.7687	0.7521	0.7355	0.7192	0.7031	0.6874	0.6720	0.6570	0.6425	0.6283	0.6147	0.6015	0.5887	0.5764
7	0.7707	0.7514	0.7320	0.7128	0.6938	0.6752	0.6570	0.6393	0.6221	0.6055	0.5895	0.5740	0.5592	0.5449	0.5312
8	0.7578	0.7359	0.7140	0.6921	0.6705	0.6494	0.6288	0.6039	0.5897	0.5712	0.5534	0.5364	0.5201	0.5046	0.4898
9	0.7464	0.7220	0.6974	0.6729	0.6488	0.6253	0.6024	0.5804	0.5593	0.5391	0.5198	0.5015	0.4840	0.4674	0.4517
10	0.7362	0.7093	0.6821	0.6550	0.6284	0.6025	0.5776	0.5536	0.5308	0.5090	0.4884	0.4689	0.4505	0.4330	0.4166

P=0.65
K=0.90

YEAR	0	5	10	15	20	25	30	35	40	45	50	55	60	65	70
0	1.0000	1.0000	1.0000	1.0000	1.0000	1.0000	1.0000	1.0000	1.0000	1.0000	1.0000	1.0000	1.0000	1.0000	1.0000
1	0.9636	0.9622	0.9609	0.9595	0.9582	0.9569	0.9557	0.9545	0.9532	0.9521	0.9509	0.9497	0.9486	0.9475	0.9464
2	0.9402	0.9373	0.9344	0.9315	0.9287	0.9260	0.9232	0.9205	0.9179	0.9153	0.9127	0.9101	0.9076	0.9052	0.9028
3	0.9231	0.9186	0.9141	0.9096	0.9052	0.9009	0.8966	0.8923	0.8881	0.8840	0.8799	0.8759	0.8720	0.8681	0.8643
4	0.9096	0.9035	0.8974	0.8913	0.8853	0.8793	0.8734	0.8676	0.8618	0.8561	0.8505	0.8450	0.8396	0.8342	0.8290
5	0.8985	0.8908	0.8831	0.8754	0.8677	0.8601	0.8525	0.8450	0.8376	0.8304	0.8232	0.8162	0.8093	0.8026	0.7960
6	0.8890	0.8798	0.8705	0.8611	0.8517	0.8424	0.8332	0.8240	0.8151	0.8062	0.7975	0.7891	0.7808	0.7726	0.7647
7	0.8809	0.8701	0.8591	0.8481	0.8370	0.8260	0.8150	0.8043	0.7937	0.7833	0.7731	0.7532	0.7535	0.7441	0.7349
8	0.8737	0.8613	0.8487	0.8360	0.8232	0.8104	0.7978	0.7854	0.7732	0.7613	0.7497	0.7384	0.7274	0.7168	0.7064
9	0.8673	0.8533	0.8391	0.8246	0.8101	0.7956	0.7814	0.7673	0.7536	0.7402	0.7272	0.7146	0.7024	0.6906	0.6791
10	0.8615	0.8460	0.8300	0.8138	0.7976	0.7814	0.7655	0.7499	0.7346	0.7198	0.7055	0.6916	0.6783	0.6654	0.6529

P=0.65
K=0.95

GROWTH

YEAR	0	5	10	15	20	25	30	35	40	45	50	55	60	65	70
0	1.0000	1.0000	1.0000	1.0000	1.0000	1.0000	1.0000	1.0000	1.0000	1.0000	1.0000	1.0000	1.0000	1.0000	1.0000
1	0.7611	0.7531	0.7454	0.7379	0.7307	0.7236	0.7168	0.7101	0.7036	0.6973	0.6912	0.6352	0.6793	0.6736	0.6681
2	0.6373	0.6232	0.6096	0.5964	0.5836	0.5713	0.5593	0.5478	0.5367	0.5259	0.5155	0.5054	0.4956	0.4862	0.4771
3	0.5587	0.5396	0.5211	0.5034	0.4864	0.4701	0.4544	0.4394	0.4251	0.4113	0.3982	0.3856	0.3735	0.3620	0.3509
4	0.5031	0.4797	0.4574	0.4360	0.4156	0.3962	0.3778	0.3604	0.3438	0.3282	0.3134	0.2995	0.2363	0.2738	0.2621
5	0.4612	0.4341	0.4083	0.3839	0.3608	0.3391	0.3187	0.2996	0.2817	0.2651	0.2495	0.2350	0.2215	0.2090	0.1973
6	0.4231	0.3977	0.3690	0.3420	0.3167	0.2932	0.2714	0.2513	0.2327	0.2157	0.2000	0.1856	0.1724	0.1603	0.1492
7	0.4012	0.3678	0.3365	0.3072	0.2802	0.2554	0.2327	0.2121	0.1933	0.1764	0.1610	0.1472	0.1346	0.1233	0.1131
8	0.3787	0.3425	0.3089	0.2778	0.2494	0.2237	0.2005	0.1797	0.1612	0.1447	0.1300	0.1169	0.1054	0.0951	0.0859
9	0.3595	0.3209	0.2851	0.2525	0.2230	0.1967	0.1733	0.1528	0.1347	0.1189	0.1051	0.0931	0.0826	0.0734	0.0653
10	0.3430	0.3020	0.2643	0.2303	0.2001	0.1735	0.1503	0.1302	0.1128	0.0979	0.0851	0.0742	0.0647	0.0566	0.0497

P=0.70
K=0.70

YEAR	0	5	10	15	20	25	30	35	40	45	50	55	60	65	70
0	1.0000	1.0000	1.0000	1.0000	1.0000	1.0000	1.0000	1.0000	1.0000	1.0000	1.0000	1.0000	1.0000	1.0000	1.0000
1	0.8023	0.7956	0.7890	0.7826	0.7764	0.7704	0.7645	0.7587	0.7531	0.7477	0.7424	0.7372	0.7321	0.7271	0.7223
2	0.6953	0.6829	0.6708	0.6591	0.6477	0.6366	0.6259	0.6154	0.6053	0.5955	0.5860	0.5767	0.5677	0.5590	0.5505
3	0.6253	0.6079	0.5912	0.5749	0.5592	0.5440	0.5293	0.5152	0.5016	0.4884	0.4758	0.4636	0.4519	0.4406	0.4297
4	0.5746	0.5530	0.5321	0.5119	0.4925	0.4739	0.4561	0.4390	0.4227	0.4071	0.3923	0.3781	0.3646	0.3518	0.3395
5	0.5357	0.5102	0.4855	0.4620	0.4394	0.4180	0.3976	0.3782	0.3600	0.3427	0.3264	0.3110	0.2965	0.2829	0.2700
6	0.5045	0.4754	0.4475	0.4209	0.3956	0.3717	0.3493	0.3282	0.3086	0.2902	0.2731	0.2571	0.2422	0.2284	0.2156
7	0.4787	0.4463	0.4154	0.3860	0.3584	0.3326	0.3085	0.2863	0.2657	0.2467	0.2292	0.2132	0.1984	0.1849	0.1724
8	0.4569	0.4215	0.3877	0.3559	0.3263	0.2988	0.2736	0.2505	0.2294	0.2103	0.1929	0.1771	0.1628	0.1499	0.1381
9	0.4382	0.3998	0.3635	0.3295	0.2981	0.2694	0.2433	0.2197	0.1986	0.1796	0.1625	0.1473	0.1337	0.1216	0.1107
10	0.4219	0.3807	0.3419	0.3060	0.2731	0.2434	0.2168	0.1931	0.1721	0.1535	0.1371	0.1227	0.1099	0.0987	0.0888

P=0.70
K=0.75

* GROWTH

YEAR*	0	5	10	15	20	25	30	35	40	45	50	55	60	65	70
0	1.0000	1.0000	1.0000	1.0000	1.0000	1.0000	1.0000	1.0000	1.0000	1.0000	1.0000	1.0000	1.0000	1.0000	1.0000
1	0.8430	0.8375	0.8321	0.8269	0.8218	0.8168	0.8119	0.8072	0.8026	0.7981	0.7937	0.7894	0.7851	0.7810	0.7770
2	0.7544	0.7439	0.7337	0.7237	0.7140	0.7045	0.6952	0.6862	0.6775	0.6689	0.6606	0.6525	0.6446	0.6369	0.6294
3	0.6947	0.6798	0.6651	0.6509	0.6371	0.6236	0.6105	0.5978	0.5855	0.5736	0.5621	0.5509	0.5400	0.5295	0.5194
4	0.6507	0.6316	0.6130	0.5949	0.5773	0.5603	0.5439	0.5281	0.5128	0.4981	0.4839	0.4703	0.4573	0.4447	0.4326
5	0.6162	0.5933	0.5710	0.5494	0.5285	0.5083	0.4890	0.4704	0.4527	0.4357	0.4196	0.4042	0.3895	0.3755	0.3622
6	0.5832	0.5517	0.5359	0.5110	0.4871	0.4641	0.4423	0.4214	0.4017	0.3830	0.3654	0.3487	0.3330	0.3181	0.3042
7	0.5547	0.5349	0.5059	0.4779	0.4512	0.4257	0.4017	0.3790	0.3577	0.3377	0.3190	0.3015	0.2852	0.2700	0.2558
8	0.5447	0.5116	0.4795	0.4487	0.4195	0.3918	0.3659	0.3417	0.3192	0.2984	0.2790	0.2512	0.2447	0.2294	0.2153
9	0.5273	0.4911	0.4561	0.4227	0.3911	0.3615	0.3341	0.3087	0.2854	0.2639	0.2443	0.2264	0.2100	0.1951	0.1814
10	0.5120	0.4728	0.4350	0.3991	0.3654	0.3342	0.3055	0.2793	0.2554	0.2337	0.2141	0.1964	0.1804	0.1659	0.1528

P=0.70
K=0.80

YEAR	0	5	10	15	20	25	30	35	40	45	50	55	60	65	70
0	1.0000	1.0000	1.0000	1.0000	1.0000	1.0000	1.0000	1.0000	1.0000	1.0000	1.0000	1.0000	1.0000	1.0000	1.0000
1	0.8830	0.8788	0.8747	0.8707	0.8668	0.8630	0.8592	0.8556	0.8520	0.8485	0.8451	0.8417	0.8385	0.8353	0.8321
2	0.8144	0.8062	0.7981	0.7901	0.7824	0.7748	0.7674	0.7602	0.7531	0.7461	0.7394	0.7328	0.7263	0.7200	0.7138
3	0.7670	0.7549	0.7431	0.7315	0.7201	0.7090	0.6981	0.6875	0.6772	0.6671	0.6573	0.6477	0.6384	0.6294	0.6206
4	0.7312	0.7156	0.7002	0.6850	0.6703	0.6558	0.6418	0.6281	0.6148	0.6019	0.5894	0.5773	0.5656	0.5542	0.5432
5	0.7028	0.6837	0.6649	0.6465	0.6284	0.6109	0.5939	0.5774	0.5615	0.5461	0.5312	0.5170	0.5032	0.4900	0.4773
6	0.6794	0.6570	0.6349	0.6133	0.5922	0.5718	0.5520	0.5330	0.5147	0.4971	0.4803	0.4642	0.4489	0.4342	0.4203
7	0.6596	0.6340	0.6088	0.5841	0.5601	0.5369	0.5146	0.4933	0.4729	0.4535	0.4351	0.4176	0.4011	0.3854	0.3705
8	0.6425	0.6138	0.5855	0.5579	0.5311	0.5054	0.4809	0.4575	0.4353	0.4144	0.3947	0.3761	0.3587	0.3422	0.3268
9	0.6275	0.5958	0.5645	0.5341	0.5047	0.4766	0.4500	0.4248	0.4012	0.3790	0.3583	0.3390	0.3209	0.3041	0.2884
10	0.6141	0.5795	0.5454	0.5122	0.4804	0.4501	0.4216	0.3949	0.3700	0.3469	0.3255	0.3056	0.2873	0.2703	0.2546

P=0.70
K=0.85

*
*
* GROWTH
*

YEAR*	0	5	10	15	20	25	30	35	40	45	50	55	60	65	70
0	1.0000	1.0000	1.0000	1.0000	1.0000	1.0000	1.0000	1.0000	1.0000	1.0000	1.0000	1.0000	1.0000	1.0000	1.0000
1	0.9225	0.9197	0.9169	0.9141	0.9115	0.9089	0.9063	0.9038	0.9014	0.8990	0.8966	0.8943	0.8921	0.8898	0.8877
2	0.3754	0.8696	0.8640	0.8584	0.8529	0.8476	0.8423	0.8371	0.8321	0.3271	0.8222	0.3174	0.8127	0.8081	0.8036
3	0.3420	0.8334	0.8249	0.8165	0.8082	0.8001	0.7922	0.7844	0.7767	0.7692	0.7618	0.7546	0.7476	0.7407	0.7339
4	0.3163	0.3030	0.7937	0.7825	0.7715	0.7607	0.7501	0.7397	0.7295	0.7196	0.7098	0.7004	0.6911	0.6821	0.6733
5	0.7956	0.7815	0.7675	0.7537	0.7400	0.7265	0.7133	0.7004	0.6878	0.6756	0.6636	0.6520	0.6407	0.6297	0.6191
6	0.7783	0.7616	0.7449	0.7284	0.7120	0.6960	0.6803	0.6650	0.6501	0.6356	0.6216	0.6081	0.5950	0.5823	0.5701
7	0.7635	0.7442	0.7249	0.7057	0.6867	0.6682	0.6501	0.6325	0.6154	0.5989	0.5831	0.5678	0.5530	0.5389	0.5253
8	0.7506	0.7287	0.7068	0.6850	0.6635	0.6425	0.6221	0.6023	0.5832	0.5649	0.5473	0.5305	0.5144	0.4990	0.4843
9	0.7392	0.7148	0.6903	0.6659	0.6419	0.6186	0.5959	0.5741	0.5532	0.5332	0.5141	0.4959	0.4786	0.4622	0.4466
10	0.7290	0.7021	0.6750	0.6481	0.6217	0.5960	0.5713	0.5475	0.5249	0.5034	0.4830	0.4637	0.4455	0.4282	0.4119

P=0.70
K=0.90

YEAR*	0	5	10	15	20	25	30	35	40	45	50	55	60	65	70
0	1.0000	1.0000	1.0000	1.0000	1.0000	1.0000	1.0000	1.0000	1.0000	1.0000	1.0000	1.0000	1.0000	1.0000	1.0000
1	0.9615	0.9600	0.9586	0.9572	0.9559	0.9545	0.9532	0.9520	0.9507	0.9495	0.9483	0.9471	0.9459	0.9448	0.9436
2	0.9373	0.9343	0.9313	0.9284	0.9255	0.9226	0.9198	0.9171	0.9144	0.9117	0.9091	0.9065	0.9040	0.9015	0.3990
3	0.9197	0.9151	0.9105	0.9060	0.9015	0.8971	0.8928	0.8885	0.8842	0.8801	0.8760	0.8719	0.8679	0.8640	0.8602
4	0.9059	0.8998	0.8936	0.8875	0.8814	0.8753	0.8694	0.8635	0.8577	0.8520	0.8463	0.8408	0.8354	0.8300	0.8248
5	0.3947	0.8869	0.8791	0.8714	0.8636	0.8560	0.8484	0.8408	0.8335	0.8262	0.8190	0.8120	0.8051	0.7984	0.7918
6	0.3851	0.3758	0.8664	0.8570	0.8476	0.8382	0.8290	0.8199	0.8109	0.8020	0.7934	0.7849	0.7766	0.7685	0.7606
7	0.3769	0.8660	0.8550	0.8439	0.8328	0.8218	0.8109	0.8001	0.7895	0.7792	0.7690	0.7591	0.7495	0.7401	0.7310
8	0.3697	0.3572	0.8445	0.8318	0.8190	0.8063	0.7937	0.7813	0.7691	0.7573	0.7457	0.7345	0.7235	0.7129	0.7026
9	0.8632	0.8492	0.8349	0.8204	0.8059	0.7915	0.7772	0.7632	0.7496	0.7362	0.7233	0.7107	0.6986	0.6868	0.6754
10	0.8574	0.8418	0.8259	0.8097	0.7934	0.7773	0.7614	0.7459	0.7307	0.7160	0.7017	0.6879	0.6746	0.6617	0.6494

P=0.70
K=0.95

* GROWTH

YEAR*	0	5	10	15	20	25	30	35	40	45	50	55	60	65	70
0	1.0000	1.0000	1.0000	1.0000	1.0000	1.0000	1.0000	1.0000	1.0000	1.0000	1.0000	1.0000	1.0000	1.0000	1.0000
1	0.7498	0.7417	0.7338	0.7261	0.7137	0.7115	0.7045	0.6978	0.6912	0.6847	0.6785	0.6724	0.6665	0.6607	0.6551
2	0.6241	0.6099	0.5962	0.5829	0.5701	0.5578	0.5459	0.5344	0.5233	0.5126	0.5023	0.4923	0.4826	0.4733	0.4642
3	0.5453	0.5262	0.5079	0.4903	0.4735	0.4573	0.4419	0.4271	0.4129	0.3994	0.3865	0.3741	0.3623	0.3510	0.3402
4	0.4900	0.4669	0.4447	0.4237	0.4036	0.3846	0.3665	0.3494	0.3333	0.3180	0.3036	0.2900	0.2771	0.2650	0.2536
5	0.4485	0.4218	0.3965	0.3725	0.3499	0.3286	0.3087	0.2901	0.2727	0.2565	0.2414	0.2273	0.2142	0.2020	0.1907
6	0.4159	0.3861	0.3579	0.3315	0.3068	0.2839	0.2627	0.2431	0.2251	0.2085	0.1933	0.1794	0.1666	0.1549	0.1441
7	0.3895	0.3567	0.3261	0.2976	0.2712	0.2471	0.2251	0.2050	0.1869	0.1704	0.1556	0.1421	0.1300	0.1191	0.1092
8	0.3674	0.3321	0.2992	0.2689	0.2413	0.2163	0.1938	0.1737	0.1557	0.1397	0.1255	0.1129	0.1017	0.0918	0.0829
9	0.3486	0.3109	0.2760	0.2442	0.2156	0.1901	0.1675	0.1476	0.1301	0.1149	0.1015	0.0899	0.0797	0.0708	0.0630
10	0.3325	0.2925	0.2558	0.2227	0.1934	0.1676	0.1451	0.1257	0.1090	0.0946	0.0822	0.0716	0.0625	0.0547	0.0479

P=0.75
K=0.70

YEAR*	0	5	10	15	20	25	30	35	40	45	50	55	60	65	70
0	1.0000	1.0000	1.0000	1.0000	1.0000	1.0000	1.0000	1.0000	1.0000	1.0000	1.0000	1.0000	1.0000	1.0000	1.0000
1	0.7927	0.7858	0.7791	0.7725	0.7661	0.7599	0.7539	0.7481	0.7424	0.7368	0.7314	0.7261	0.7209	0.7159	0.7110
2	0.6837	0.6711	0.6589	0.6471	0.6356	0.6245	0.6137	0.6033	0.5932	0.5833	0.5738	0.5646	0.5556	0.5470	0.5385
3	0.6131	0.5958	0.5790	0.5628	0.5471	0.5320	0.5175	0.5035	0.4900	0.4770	0.4645	0.4525	0.4409	0.4298	0.4191
4	0.5625	0.5410	0.5202	0.5002	0.4810	0.4627	0.4451	0.4283	0.4122	0.3969	0.3823	0.3684	0.3552	0.3426	0.3306
5	0.5238	0.4985	0.4742	0.4509	0.4287	0.4076	0.3875	0.3686	0.3506	0.3337	0.3178	0.3027	0.2886	0.2753	0.2627
6	0.4929	0.4641	0.4365	0.4104	0.3856	0.3622	0.3402	0.3196	0.3003	0.2824	0.2657	0.2501	0.2356	0.2222	0.2096
7	0.4674	0.4355	0.4050	0.3762	0.3491	0.3238	0.3003	0.2786	0.2585	0.2400	0.2229	0.2073	0.1929	0.1798	0.1676
8	0.4459	0.4110	0.3778	0.3467	0.3177	0.2908	0.2662	0.2437	0.2231	0.2045	0.1875	0.1722	0.1583	0.1457	0.1343
9	0.4275	0.3897	0.3541	0.3208	0.2901	0.2621	0.2366	0.2137	0.1931	0.1746	0.1580	0.1432	0.1300	0.1182	0.1076
10	0.4114	0.3710	0.3330	0.2978	0.2657	0.2368	0.2108	0.1877	0.1673	0.1492	0.1333	0.1192	0.1068	0.0959	0.0863

P=0.75
K=0.75

*
 *
  * GROWTH
   *

YEAR*	0	5	10	15	20	25	30	35	40	45	50	55	60	65	70	
0	1.0000	1.0000	1.0000	1.0000	1.0000	1.0000	1.0000	1.0000	1.0000	1.0000	1.0000	1.0000	1.0000	1.0000	1.0000	
1	0.3351	0.3295	0.8239	0.8186	0.8133	0.8082	0.8032	0.7984	0.7937	0.7890	0.7845	0.7301	0.7758	0.7716	0.7675	
2	0.7445	0.7339	0.7235	0.7134	0.7036	0.6941	0.6848	0.6757	0.6669	0.6583	0.6500	0.6418	0.6339	0.6262	0.6187	
3	0.6342	0.6692	0.6545	0.6403	0.6264	0.6130	0.5999	0.5873	0.5750	0.5632	0.5517	0.5406	0.5299	0.5195	0.5094	********
4	0.6400	0.6209	0.6024	0.5843	0.5669	0.5500	0.5337	0.5180	0.5029	0.4883	0.4744	0.4510	0.4481	0.4357	0.4238	*P=0.75*
5	0.6056	0.5828	0.5606	0.5391	0.5184	0.4985	0.4794	0.4611	0.4436	0.4269	0.4110	0.3958	0.3814	0.3677	0.3546	********
6	0.5776	0.5513	0.5258	0.5012	0.4775	0.4548	0.4333	0.4123	0.3934	0.3750	0.3577	0.3413	0.3259	0.3113	0.2976	*K=0.80*
7	0.5544	0.5247	0.4960	0.4684	0.4421	0.4170	0.3934	0.3711	0.3501	0.3305	0.3122	0.2951	0.2791	0.2642	0.2503	********
8	0.5345	0.5017	0.4700	0.4397	0.4109	0.3837	0.3582	0.3345	0.3124	0.2919	0.2730	0.2555	0.2393	0.2244	0.2107	
9	0.5173	0.4815	0.4469	0.4140	0.3829	0.3539	0.3270	0.3021	0.2792	0.2582	0.2390	0.2215	0.2054	0.1908	0.1774	
10	0.5021	0.4634	0.4261	0.3908	0.3577	0.3271	0.2990	0.2732	0.2498	0.2286	0.2095	0.1921	0.1764	0.1623	0.1495	
0	1.0000	1.0000	1.0000	1.0000	1.0000	1.0000	1.0000	1.0000	1.0000	1.0000	1.0000	1.0000	1.0000	1.0000	1.0000	
1	0.8770	0.8727	0.8684	0.8643	0.8603	0.8564	0.8525	0.8488	0.8451	0.8415	0.8380	0.8346	0.8312	0.8279	0.8247	
2	0.8067	0.7983	0.7900	0.7820	0.7741	0.7665	0.7590	0.7515	0.7445	0.7375	0.7307	0.7240	0.7175	0.7112	0.7049	
3	0.7585	0.7464	0.7344	0.7227	0.7113	0.7001	0.6892	0.6786	0.6683	0.6582	0.6485	0.6389	0.6296	0.6206	0.6119	********
4	0.7225	0.7068	0.6913	0.6762	0.6614	0.6470	0.6330	0.6194	0.6061	0.5933	0.5809	0.5689	0.5573	0.5460	0.5351	*P=0.75*
5	0.6940	0.6748	0.6560	0.6376	0.6197	0.6023	0.5853	0.5690	0.5532	0.5379	0.5233	0.5091	0.4956	0.4825	0.4700	********
6	0.6705	0.6481	0.6261	0.6046	0.5837	0.5634	0.5438	0.5250	0.5069	0.4895	0.4729	0.4571	0.4419	0.4275	0.4137	*K=0.85*
7	0.6507	0.6252	0.6001	0.5756	0.5518	0.5289	0.5068	0.4858	0.4656	0.4465	0.4283	0.4111	0.3947	0.3793	0.3646	********
8	0.6337	0.6051	0.5770	0.5496	0.5232	0.4977	0.4735	0.4504	0.4285	0.4079	0.3885	0.3702	0.3530	0.3368	0.3216	
9	0.6187	0.5872	0.5562	0.5261	0.4970	0.4693	0.4430	0.4182	0.3949	0.3730	0.3526	0.3336	0.3158	0.2993	0.2838	
10	0.6055	0.5711	0.5373	0.5045	0.4730	0.4431	0.4150	0.3887	0.3642	0.3414	0.3203	0.3007	0.2827	0.2660	0.2505	

* GROWTH

YEAR*	0	5	10	15	20	25	30	35	40	45	50	55	60	65	70
0	1.0000	1.0000	1.0000	1.0000	1.0000	1.0000	1.0000	1.0000	1.0000	1.0000	1.0000	1.0000	1.0000	1.0000	1.0000
1	0.9185	0.9155	0.9126	0.9098	0.9070	0.9044	0.9017	0.8991	0.8966	0.8942	0.8917	0.8894	0.8871	0.8848	0.8825
2	0.8700	0.8641	0.8583	0.8526	0.8471	0.8416	0.8363	0.8310	0.8259	0.8209	0.8159	0.8111	0.8064	0.8017	0.7972
3	0.8360	0.8272	0.8186	0.8102	0.8018	0.7937	0.7856	0.7778	0.7701	0.7625	0.7552	0.7479	0.7409	0.7340	0.7273
4	0.8100	0.7985	0.7871	0.7759	0.7649	0.7541	0.7434	0.7330	0.7228	0.7129	0.7032	0.6937	0.6845	0.6755	0.6663
5	0.7891	0.7749	0.7609	0.7470	0.7333	0.7198	0.7067	0.6938	0.6812	0.6690	0.6571	0.6456	0.6344	0.6235	0.6129
6	0.7717	0.7549	0.7382	0.7217	0.7054	0.6894	0.6737	0.6585	0.6437	0.6293	0.6154	0.6020	0.5890	0.5764	0.5643
7	0.7569	0.7375	0.7182	0.6990	0.6802	0.6617	0.6437	0.6262	0.6093	0.5929	0.5771	0.5620	0.5474	0.5334	0.5199
8	0.7439	0.7221	0.7001	0.6784	0.6570	0.6362	0.6159	0.5962	0.5773	0.5592	0.5417	0.5250	0.5091	0.4939	0.4793
9	0.7325	0.7081	0.6837	0.6594	0.6356	0.6124	0.5899	0.5683	0.5475	0.5277	0.5088	0.4903	0.4737	0.4574	0.4420
10	0.7223	0.6955	0.6685	0.6417	0.6155	0.5900	0.5655	0.5419	0.5195	0.4982	0.4780	0.4589	0.4408	0.4238	0.4076

P=0.75
K=0.90

YEAR*	0	5	10	15	20	25	30	35	40	45	50	55	60	65	70
0	1.0000	1.0000	1.0000	1.0000	1.0000	1.0000	1.0000	1.0000	1.0000	1.0000	1.0000	1.0000	1.0000	1.0000	1.0000
1	0.9594	0.9579	0.9565	0.9550	0.9536	0.9522	0.9509	0.9496	0.9483	0.9470	0.9457	0.9445	0.9433	0.9421	0.9410
2	0.9344	0.9314	0.9283	0.9253	0.9224	0.9195	0.9166	0.9138	0.9111	0.9084	0.9057	0.9031	0.9005	0.8980	0.8955
3	0.9165	0.9118	0.9072	0.9026	0.8981	0.8936	0.8892	0.8848	0.8806	0.8764	0.8722	0.8682	0.8642	0.8602	0.8564
4	0.9025	0.8962	0.8900	0.8838	0.8777	0.8716	0.8656	0.8597	0.8538	0.8481	0.8425	0.8369	0.8315	0.8261	0.8209
5	0.8911	0.8833	0.8754	0.8676	0.8598	0.8521	0.8445	0.8370	0.8296	0.8223	0.8151	0.8081	0.8013	0.7945	0.7880
6	0.8815	0.8721	0.8626	0.8532	0.8437	0.8344	0.8251	0.8160	0.8070	0.7982	0.7895	0.7811	0.7728	0.7647	0.7569
7	0.8732	0.8622	0.8512	0.8400	0.8289	0.8179	0.8070	0.7962	0.7857	0.7753	0.7652	0.7554	0.7457	0.7364	0.7273
8	0.8659	0.8534	0.8407	0.8279	0.8151	0.8024	0.7898	0.7774	0.7653	0.7535	0.7420	0.7308	0.7199	0.7093	0.6991
9	0.8594	0.8453	0.8310	0.8165	0.8020	0.7876	0.7734	0.7595	0.7458	0.7326	0.7197	0.7072	0.6950	0.6833	0.6720
10	0.8535	0.8379	0.8220	0.8058	0.7896	0.7735	0.7576	0.7421	0.7270	0.7123	0.6981	0.6844	0.6712	0.6584	0.6461

P=0.75
K=0.95

*
 *
  * GROWTH
   *

YEAR*	0	5	10	15	20	25	30	35	40	45	50	55	60	65	70
0	1.0000	1.0000	1.0000	1.0000	1.0000	1.0000	1.0000	1.0000	1.0000	1.0000	1.0000	1.0000	1.0000	1.0000	1.0000
1	0.7390	0.7307	0.7226	0.7149	0.7073	0.7000	0.6929	0.6860	0.6793	0.6728	0.6665	0.6603	0.6544	0.6485	0.6428
2	0.6116	0.5973	0.5835	0.5703	0.5576	0.5453	0.5334	0.5219	0.5109	0.5003	0.4900	0.4801	0.4705	0.4613	0.4524
3	0.5327	0.5133	0.4956	0.4782	0.4615	0.4455	0.4303	0.4157	0.4018	0.3885	0.3758	0.3636	0.3521	0.3410	0.3304
4	0.4779	0.4550	0.4331	0.4123	0.3926	0.3739	0.3562	0.3394	0.3236	0.3087	0.2946	0.2813	0.2688	0.2570	0.2458
5	0.4368	0.4105	0.3856	0.3620	0.3399	0.3191	0.2996	0.2814	0.2645	0.2487	0.2340	0.2203	0.2075	0.1957	0.1847
6	0.4047	0.3754	0.3477	0.3218	0.2977	0.2754	0.2547	0.2356	0.2181	0.2020	0.1873	0.1737	0.1613	0.1499	0.1395
7	0.3787	0.3466	0.3166	0.2887	0.2630	0.2395	0.2181	0.1986	0.1810	0.1650	0.1506	0.1376	0.1259	0.1153	0.1057
8	0.3570	0.3224	0.2903	0.2608	0.2339	0.2096	0.1877	0.1682	0.1508	0.1353	0.1215	0.1093	0.0984	0.0888	0.0803
9	0.3387	0.3017	0.2677	0.2368	0.2089	0.1841	0.1622	0.1429	0.1260	0.1112	0.0982	0.0870	0.0771	0.0685	0.0610
10	0.3223	0.2837	0.2480	0.2158	0.1873	0.1623	0.1405	0.1217	0.1054	0.0915	0.0795	0.0593	0.0605	0.0529	0.0464

****
P=0.80
****
K=0.70
****

YEAR*	0	5	10	15	20	25	30	35	40	45	50	55	60	65	70
0	1.0000	1.0000	1.0000	1.0000	1.0000	1.0000	1.0000	1.0000	1.0000	1.0000	1.0000	1.0000	1.0000	1.0000	1.0000
1	0.7835	0.7764	0.7695	0.7628	0.7563	0.7500	0.7439	0.7379	0.7321	0.7264	0.7209	0.7155	0.7103	0.7052	0.7002
2	0.6726	0.6599	0.6477	0.6358	0.6243	0.6131	0.6023	0.5919	0.5818	0.5720	0.5625	0.5533	0.5444	0.5358	0.5274
3	0.6018	0.5844	0.5677	0.5515	0.5359	0.5209	0.5065	0.4926	0.4793	0.4664	0.4541	0.4422	0.4308	0.4199	0.4094
4	0.5512	0.5298	0.5092	0.4894	0.4704	0.4523	0.4349	0.4183	0.4026	0.3875	0.3732	0.3596	0.3466	0.3342	0.3225
5	0.5127	0.4877	0.4636	0.4407	0.4188	0.3980	0.3783	0.3597	0.3421	0.3255	0.3099	0.2952	0.2813	0.2683	0.2561
6	0.4821	0.4537	0.4266	0.4008	0.3764	0.3534	0.3318	0.3117	0.2928	0.2753	0.2589	0.2437	0.2296	0.2164	0.2042
7	0.4569	0.4254	0.3955	0.3671	0.3406	0.3158	0.2928	0.2715	0.2519	0.2338	0.2172	0.2019	0.1879	0.1751	0.1633
8	0.4357	0.4014	0.3688	0.3382	0.3098	0.2835	0.2594	0.2374	0.2174	0.1992	0.1827	0.1677	0.1541	0.1419	0.1307
9	0.4176	0.3804	0.3454	0.3128	0.2828	0.2554	0.2306	0.2082	0.1881	0.1700	0.1539	0.1395	0.1266	0.1151	0.1048
10	0.4017	0.3620	0.3248	0.2904	0.2590	0.2307	0.2054	0.1829	0.1629	0.1453	0.1298	0.1161	0.1040	0.0934	0.0840

****
P=0.80
****
K=0.75
****

GROWTH

YEAR	0	5	10	15	20	25	30	35	40	45	50	55	60	65	70
0	1.0000	1.0000	1.0000	1.0000	1.0000	1.0000	1.0000	1.0000	1.0000	1.0000	1.0000	1.0000	1.0000	1.0000	1.0000
1	0.3276	0.8218	0.8161	0.8106	0.8052	0.8000	0.7949	0.7900	0.7851	0.7804	0.7758	0.7713	0.7670	0.7627	0.7585
2	0.7352	0.7244	0.7140	0.7038	0.6939	0.6842	0.6749	0.6658	0.6570	0.6484	0.6400	0.6319	0.6240	0.6163	0.6088
3	0.6744	0.6593	0.6445	0.6303	0.6164	0.6030	0.5900	0.5774	0.5652	0.5535	0.5421	0.5311	0.5204	0.5101	0.5002
4	0.6300	0.6110	0.5924	0.5745	0.5571	0.5404	0.5242	0.5087	0.4937	0.4793	0.4655	0.4523	0.4396	0.4274	0.4157
5	0.5956	0.5729	0.5509	0.5296	0.5091	0.4894	0.4705	0.4524	0.4351	0.4187	0.4030	0.3881	0.3739	0.3604	0.3476
6	0.5678	0.5417	0.5164	0.4920	0.4686	0.4463	0.4250	0.4048	0.3857	0.3677	0.3506	0.3345	0.3194	0.3051	0.2916
7	0.5447	0.5154	0.4870	0.4597	0.4337	0.4090	0.3857	0.3638	0.3432	0.3239	0.3059	0.2891	0.2735	0.2588	0.2452
8	0.5250	0.4926	0.4613	0.4313	0.4029	0.3762	0.3512	0.3278	0.3061	0.2861	0.2675	0.2503	0.2345	0.2199	0.2064
9	0.5079	0.4726	0.4385	0.4060	0.3755	0.3469	0.3205	0.2960	0.2736	0.2530	0.2342	0.2170	0.2013	0.1869	0.1733
10	0.4929	0.4547	0.4180	0.3832	0.3507	0.3206	0.2929	0.2677	0.2448	0.2240	0.2052	0.1882	0.1728	0.1590	0.1464

P=0.80
K=0.80

YEAR	0	5	10	15	20	25	30	35	40	45	50	55	60	65	70
0	1.0000	1.0000	1.0000	1.0000	1.0000	1.0000	1.0000	1.0000	1.0000	1.0000	1.0000	1.0000	1.0000	1.0000	1.0000
1	0.8713	0.8668	0.8624	0.8582	0.8540	0.8500	0.8461	0.8422	0.8385	0.8348	0.8312	0.8277	0.8243	0.8209	0.8176
2	0.7993	0.7907	0.7824	0.7743	0.7663	0.7585	0.7510	0.7436	0.7364	0.7294	0.7225	0.7158	0.7093	0.7029	0.6967
3	0.7506	0.7383	0.7262	0.7145	0.7030	0.6918	0.6809	0.6703	0.6600	0.6500	0.5402	0.6307	0.6215	0.6125	0.6038
4	0.7143	0.6985	0.6830	0.6679	0.6531	0.6387	0.6248	0.6112	0.5981	0.5853	0.5730	0.5611	0.5496	0.5384	0.5276
5	0.6857	0.6665	0.6478	0.6294	0.6116	0.5942	0.5774	0.5612	0.5455	0.5304	0.5159	0.5019	0.4885	0.4756	0.4632
6	0.6622	0.6399	0.6180	0.5966	0.5758	0.5556	0.5362	0.5176	0.4997	0.4825	0.4661	0.4504	0.4355	0.4212	0.4076
7	0.6425	0.6171	0.5921	0.5678	0.5442	0.5214	0.4996	0.4788	0.4589	0.4400	0.4221	0.4050	0.3889	0.3737	0.3592
8	0.6255	0.5971	0.5692	0.5420	0.5158	0.4906	0.4666	0.4438	0.4223	0.4019	0.3827	0.3647	0.3477	0.3318	0.3168
9	0.6106	0.5793	0.5485	0.5187	0.4899	0.4625	0.4366	0.4121	0.3891	0.3675	0.3474	0.3286	0.3111	0.2948	0.2796
10	0.5974	0.5633	0.5298	0.4973	0.4662	0.4367	0.4089	0.3830	0.3598	0.3363	0.3155	0.2963	0.2785	0.2620	0.2468

P=0.80
K=0.85

* GROWTH

YEAR*	0	5	10	15	20	25	30	35	40	45	50	55	60	65	70
0	1.0000	1.0000	1.0000	1.0000	1.0000	1.0000	1.0000	1.0000	1.0000	1.0000	1.0000	1.0000	1.0000	1.0000	1.0000
1	0.9145	0.9115	0.9085	0.9056	0.9028	0.9000	0.8973	0.8947	0.8921	0.8895	0.8871	0.8846	0.8823	0.8799	0.8776
2	0.8648	0.8588	0.8529	0.8472	0.8415	0.8360	0.8306	0.8253	0.8201	0.8150	0.8100	0.8051	0.8004	0.7957	0.7911
3	0.8303	0.8214	0.8127	0.8042	0.7958	0.7875	0.7795	0.7716	0.7639	0.7563	0.7489	0.7417	0.7346	0.7277	0.7210
4	0.8040	0.7924	0.7810	0.7697	0.7587	0.7478	0.7372	0.7268	0.7166	0.7067	0.6970	0.6875	0.6784	0.6694	0.6607
5	0.7830	0.7688	0.7546	0.7407	0.7270	0.7136	0.7005	0.6876	0.6751	0.6629	0.6511	0.6395	0.6285	0.6176	0.6072
6	0.7655	0.7487	0.7320	0.7154	0.6992	0.6832	0.6676	0.6525	0.6378	0.6235	0.6096	0.5963	0.5834	0.5709	0.5589
7	0.7506	0.7313	0.7119	0.6928	0.6740	0.6556	0.6377	0.6204	0.6035	0.5873	0.5717	0.5566	0.5421	0.5283	0.5149
8	0.7377	0.7158	0.6939	0.6723	0.6510	0.6303	0.6101	0.5906	0.5718	0.5538	0.5365	0.5200	0.5042	0.4891	0.4747
9	0.7263	0.7019	0.6775	0.6534	0.6297	0.6066	0.5843	0.5628	0.5423	0.5226	0.5039	0.4859	0.4691	0.4530	0.4377
10	0.7161	0.6893	0.6624	0.6358	0.6097	0.5844	0.5601	0.5367	0.5145	0.4934	0.4734	0.4544	0.4365	0.4196	0.4037

P=0.80
K=0.90

YEAR*	0	5	10	15	20	25	30	35	40	45	50	55	60	65	70
0	1.0000	1.0000	1.0000	1.0000	1.0000	1.0000	1.0000	1.0000	1.0000	1.0000	1.0000	1.0000	1.0000	1.0000	1.0000
1	0.9574	0.9559	0.9544	0.9529	0.9514	0.9500	0.9486	0.9472	0.9459	0.9446	0.9433	0.9421	0.9408	0.9396	0.9384
2	0.9317	0.9286	0.9255	0.9224	0.9194	0.9165	0.9136	0.9107	0.9079	0.9052	0.9025	0.8999	0.8973	0.8947	0.8922
3	0.9134	0.9087	0.9040	0.8993	0.8948	0.8902	0.8858	0.8814	0.8771	0.8729	0.8687	0.8646	0.8606	0.8567	0.8528
4	0.8992	0.8929	0.8866	0.8804	0.8742	0.8681	0.8620	0.8561	0.8502	0.8445	0.8388	0.8333	0.8278	0.8225	0.8173
5	0.8877	0.8798	0.8719	0.8641	0.8562	0.8485	0.8409	0.8333	0.8259	0.8186	0.8115	0.8045	0.7976	0.7909	0.7843
6	0.8780	0.8686	0.8591	0.8496	0.8401	0.8307	0.8214	0.8123	0.8033	0.7945	0.7859	0.7775	0.7692	0.7612	0.7533
7	0.8697	0.8587	0.8475	0.8364	0.8253	0.8142	0.8033	0.7926	0.7821	0.7717	0.7617	0.7518	0.7423	0.7329	0.7239
8	0.8623	0.8498	0.8370	0.8242	0.8114	0.7987	0.7862	0.7739	0.7618	0.7500	0.7385	0.7273	0.7165	0.7060	0.6957
9	0.8558	0.8417	0.8274	0.8129	0.7984	0.7840	0.7698	0.7559	0.7423	0.7291	0.7163	0.7038	0.6918	0.6801	0.6688
10	0.8499	0.8343	0.8183	0.8021	0.7859	0.7699	0.7541	0.7387	0.7236	0.7090	0.6948	0.6812	0.6680	0.6552	0.6430

P=0.80
K=0.95

*

* GROWTH

*

YEAR*	0	5	10	15	20	25	30	35	40	45	50	55	60	65	70
0	1.0000	1.0000	1.0000	1.0000	1.0000	1.0000	1.0000	1.0000	1.0000	1.0000	1.0000	1.0000	1.0000	1.0000	1.0000
1	0.7237	0.7202	0.7120	0.7041	0.6964	0.6890	0.6818	0.6748	0.6681	0.6615	0.6551	0.6489	0.6428	0.6370	0.6313
2	0.5998	0.5855	0.5718	0.5585	0.5458	0.5335	0.5217	0.5103	0.4993	0.4888	0.4786	0.4588	0.4593	0.4502	0.4413
3	0.5210	0.5022	0.4841	0.4669	0.4503	0.4346	0.4195	0.4052	0.3914	0.3784	0.3659	0.3540	0.3426	0.3318	0.3214
4	0.4665	0.4439	0.4223	0.4018	0.3824	0.3640	0.3467	0.3302	0.3148	0.3002	0.2664	0.2734	0.2612	0.2496	0.2387
5	0.4260	0.4001	0.3755	0.3524	0.3307	0.3103	0.2913	0.2735	0.2569	0.2415	0.2272	0.2139	0.2015	0.1899	0.1792
6	0.3944	0.3655	0.3384	0.3130	0.2894	0.2676	0.2474	0.2288	0.2117	0.1961	0.1817	0.1686	0.1565	0.1454	0.1353
7	0.3688	0.3373	0.3079	0.2806	0.2555	0.2326	0.2117	0.1928	0.1756	0.1601	0.1461	0.1335	0.1221	0.1118	0.1025
8	0.3475	0.3136	0.2822	0.2533	0.2271	0.2034	0.1822	0.1632	0.1462	0.1312	0.1178	0.1060	0.0955	0.0861	0.0778
9	0.3295	0.2933	0.2601	0.2299	0.2028	0.1787	0.1573	0.1386	0.1222	0.1078	0.0952	0.0843	0.0748	0.0664	0.0591
10	0.3140	0.2758	0.2409	0.2095	0.1818	0.1574	0.1363	0.1180	0.1022	0.0887	0.0771	0.0671	0.0596	0.0513	0.0450

P=0.85

K=0.70

YEAR*	0	5	10	15	20	25	30	35	40	45	50	55	60	65	70
0	1.0000	1.0000	1.0000	1.0000	1.0000	1.0000	1.0000	1.0000	1.0000	1.0000	1.0000	1.0000	1.0000	1.0000	1.0000
1	0.7747	0.7674	0.7604	0.7535	0.7469	0.7405	0.7342	0.7282	0.7223	0.7165	0.7110	0.7055	0.7002	0.6950	0.6900
2	0.6622	0.6494	0.6371	0.6251	0.6136	0.6024	0.5917	0.5812	0.5711	0.5614	0.5519	0.5428	0.5339	0.5253	0.5170
3	0.5911	0.5738	0.5571	0.5410	0.5255	0.5106	0.4963	0.4825	0.4693	0.4566	0.4444	0.4327	0.4215	0.4107	0.4003
4	0.5407	0.5194	0.4990	0.4793	0.4606	0.4426	0.4255	0.4092	0.3936	0.3788	0.3647	0.3513	0.3386	0.3265	0.3150
5	0.5025	0.4776	0.4539	0.4312	0.4096	0.3891	0.3697	0.3514	0.3342	0.3179	0.3026	0.2882	0.2747	0.2619	0.2499
6	0.4721	0.4441	0.4173	0.3919	0.3679	0.3453	0.3241	0.3044	0.2859	0.2687	0.2527	0.2379	0.2240	0.2112	0.1992
7	0.4472	0.4152	0.3866	0.3588	0.3327	0.3084	0.2859	0.2651	0.2459	0.2282	0.2119	0.1970	0.1833	0.1708	0.1593
8	0.4263	0.3924	0.3604	0.3304	0.3025	0.2768	0.2532	0.2317	0.2121	0.1943	0.1782	0.1636	0.1504	0.1384	0.1275
9	0.4084	0.3719	0.3375	0.3055	0.2761	0.2493	0.2250	0.2031	0.1835	0.1658	0.1501	0.1360	0.1235	0.1122	0.1022
10	0.3928	0.3538	0.3172	0.2835	0.2528	0.2251	0.2004	0.1784	0.1589	0.1417	0.1266	0.1132	0.1014	0.0911	0.0819

P=0.85

K=0.75

* GROWTH

YEAR*	0	5	10	15	20	25	30	35	40	45	50	55	60	65	70
0	1.0000	1.0000	1.0000	1.0000	1.0000	1.0000	1.0000	1.0000	1.0000	1.0000	1.0000	1.0000	1.0000	1.0000	1.0000
1	0.8203	0.8144	0.8086	0.8029	0.7974	0.7921	0.7869	0.7819	0.7770	0.7722	0.7675	0.7629	0.7585	0.7541	0.7499
2	0.7263	0.7154	0.7049	0.6946	0.6847	0.6750	0.6656	0.6565	0.6476	0.6390	0.6306	0.6225	0.6146	0.6069	0.5995
3	0.6651	0.6499	0.6352	0.6209	0.6071	0.5937	0.5808	0.5682	0.5561	0.5444	0.5331	0.5222	0.5116	0.5014	0.4916
4	0.6207	0.6016	0.5832	0.5653	0.5481	0.5314	0.5154	0.5000	0.4852	0.4710	0.4574	0.4443	0.4317	0.4197	0.4082
5	0.5864	0.5638	0.5419	0.5207	0.5004	0.4809	0.4622	0.4444	0.4273	0.4111	0.3957	0.3810	0.3670	0.3537	0.3411
6	0.5587	0.5328	0.5077	0.4835	0.4604	0.4383	0.4173	0.3975	0.3786	0.3609	0.3441	0.3283	0.3134	0.2993	0.2861
7	0.5357	0.5066	0.4785	0.4516	0.4259	0.4016	0.3786	0.3570	0.3368	0.3179	0.3002	0.2837	0.2683	0.2539	0.2405
8	0.5162	0.4841	0.4531	0.4236	0.3956	0.3692	0.3446	0.3217	0.3004	0.2806	0.2624	0.2456	0.2300	0.2156	0.2024
9	0.4993	0.4643	0.4305	0.3986	0.3685	0.3405	0.3144	0.2904	0.2684	0.2482	0.2297	0.2128	0.1974	0.1833	0.1704
10	0.4844	0.4467	0.4104	0.3762	0.3442	0.3146	0.2874	0.2626	0.2401	0.2197	0.2012	0.1846	0.1695	0.1559	0.1436

P=0.85
K=0.80

YEAR*	0	5	10	15	20	25	30	35	40	45	50	55	60	65	70
0	1.0000	1.0000	1.0000	1.0000	1.0000	1.0000	1.0000	1.0000	1.0000	1.0000	1.0000	1.0000	1.0000	1.0000	1.0000
1	0.3657	0.3611	0.8566	0.8523	0.8480	0.8439	0.8399	0.8359	0.8321	0.8284	0.8247	0.8211	0.8176	0.8142	0.8109
2	0.7922	0.7836	0.7751	0.7669	0.7589	0.7510	0.7434	0.7360	0.7287	0.7217	0.7148	0.7081	0.7015	0.6951	0.6889
3	0.7430	0.7306	0.7185	0.7067	0.6952	0.6840	0.6731	0.6625	0.6522	0.6422	0.6325	0.6230	0.6138	0.6049	0.5962
4	0.7065	0.6907	0.6752	0.6601	0.6453	0.6310	0.6171	0.6036	0.5905	0.5779	0.5657	0.5538	0.5424	0.5313	0.5207
5	0.6779	0.6587	0.6400	0.6217	0.6040	0.5867	0.5700	0.5539	0.5384	0.5234	0.5090	0.4952	0.4319	0.4691	0.4569
6	0.6544	0.6322	0.6103	0.5890	0.5684	0.5484	0.5292	0.5107	0.4930	0.4760	0.4598	0.4443	0.4295	0.4154	0.4020
7	0.6347	0.6094	0.5845	0.5604	0.5370	0.5145	0.4929	0.4723	0.4527	0.4340	0.4163	0.3994	0.3835	0.3685	0.3542
8	0.6178	0.5895	0.5613	0.5349	0.5089	0.4840	0.4603	0.4378	0.4165	0.3964	0.3774	0.3596	0.3429	0.3272	0.3124
9	0.6030	0.5719	0.5414	0.5118	0.4833	0.4562	0.4305	0.4064	0.3337	0.3624	0.3425	0.3240	0.3067	0.2906	0.2756
10	0.5899	0.5560	0.5223	0.4906	0.4598	0.4307	0.4033	0.3776	0.3538	0.3316	0.3111	0.2921	0.2745	0.2583	0.2433

P=0.85
K=0.85

* GROWTH

YEAR*	0	5	10	15	20	25	30	35	40	45	50	55	60	65	70
0	1.0000	1.0000	1.0000	1.0000	1.0000	1.0000	1.0000	1.0000	1.0000	1.0000	1.0000	1.0000	1.0000	1.0000	1.0000
1	0.9107	0.9076	0.9045	0.9015	0.8986	0.8958	0.8930	0.8903	0.8877	0.8851	0.8825	0.8801	0.8776	0.8753	0.8729
2	0.8599	0.8538	0.8478	0.8419	0.8362	0.8306	0.8251	0.8198	0.8145	0.8094	0.8044	0.7995	0.7947	0.7900	0.7854
3	0.8248	0.8159	0.8071	0.7985	0.7901	0.7818	0.7737	0.7658	0.7580	0.7504	0.7430	0.7358	0.7288	0.7219	0.7151
4	0.7983	0.7867	0.7752	0.7639	0.7528	0.7419	0.7313	0.7209	0.7107	0.7008	0.6912	0.6818	0.6726	0.6637	0.6550
5	0.7772	0.7629	0.7488	0.7348	0.7212	0.7077	0.6946	0.6818	0.6694	0.6573	0.6455	0.6340	0.6230	0.6122	0.6018
6	0.7597	0.7428	0.7261	0.7096	0.6933	0.6774	0.6619	0.6468	0.6322	0.6180	0.6043	0.5910	0.5782	0.5658	0.5539
7	0.7448	0.7254	0.7061	0.6870	0.6683	0.6500	0.6322	0.6149	0.5982	0.5821	0.5665	0.5516	0.5373	0.5235	0.5103
8	0.7318	0.7099	0.6881	0.6666	0.6454	0.6247	0.6047	0.5854	0.5667	0.5488	0.5317	0.5153	0.4996	0.4846	0.4703
9	0.7204	0.6961	0.6718	0.6477	0.6242	0.6013	0.5791	0.5578	0.5374	0.5179	0.4993	0.4816	0.4648	0.4488	0.4337
10	0.7102	0.6835	0.6567	0.6302	0.6043	0.5792	0.5550	0.5319	0.5098	0.4889	0.4691	0.4503	0.4326	0.4158	0.4000

P=0.85
K=0.90

YEAR*	0	5	10	15	20	25	30	35	40	45	50	55	60	65	70
0	1.0000	1.0000	1.0000	1.0000	1.0000	1.0000	1.0000	1.0000	1.0000	1.0000	1.0000	1.0000	1.0000	1.0000	1.0000
1	0.9555	0.9539	0.9523	0.9508	0.9493	0.9478	0.9464	0.9450	0.9436	0.9423	0.9410	0.9397	0.9384	0.9372	0.9360
2	0.9291	0.9259	0.9228	0.9196	0.9166	0.9136	0.9107	0.9078	0.9050	0.9022	0.8994	0.8968	0.8941	0.8916	0.8890
3	0.9105	0.9057	0.9009	0.8962	0.8916	0.8871	0.8826	0.8782	0.8738	0.8696	0.8654	0.8613	0.8572	0.8533	0.8494
4	0.8962	0.8898	0.8834	0.8771	0.8709	0.8647	0.8587	0.8527	0.8468	0.8411	0.8354	0.8299	0.8244	0.8191	0.8138
5	0.8845	0.8766	0.8686	0.8607	0.8529	0.8451	0.8374	0.8299	0.8225	0.8152	0.8081	0.8011	0.7942	0.7875	0.7810
6	0.8748	0.8652	0.8557	0.8462	0.8367	0.8273	0.8180	0.8089	0.7999	0.7911	0.7825	0.7741	0.7659	0.7579	0.7500
7	0.8663	0.8553	0.8441	0.8330	0.8218	0.8108	0.7999	0.7892	0.7787	0.7684	0.7583	0.7485	0.7390	0.7297	0.7207
8	0.8590	0.8464	0.8336	0.8208	0.8080	0.7953	0.7828	0.7705	0.7585	0.7467	0.7353	0.7241	0.7133	0.7028	0.6927
9	0.8524	0.8383	0.8239	0.8094	0.7950	0.7806	0.7665	0.7526	0.7391	0.7259	0.7131	0.7007	0.6887	0.6771	0.6658
10	0.8465	0.8309	0.8149	0.7987	0.7826	0.7665	0.7508	0.7354	0.7204	0.7058	0.6917	0.6731	0.6650	0.6523	0.6401

P=0.85
K=0.95

GROWTH

ρ = 0.90, K = 0.70

YEAR	0	5	10	15	20	25	30	35	40	45	50	55	60	65	70
0	1.0000	1.0000	1.0000	1.0000	1.0000	1.0000	1.0000	1.0000	1.0000	1.0000	1.0000	1.0000	1.0000	1.0000	1.0000
1	0.7187	0.7101	0.7018	0.6938	0.6860	0.6785	0.6712	0.6642	0.6573	0.6507	0.6443	0.6380	0.6319	0.6260	0.6202
2	0.5887	0.5744	0.5606	0.5474	0.5347	0.5225	0.5107	0.4994	0.4885	0.4780	0.4679	0.4582	0.4488	0.4398	0.4311
3	0.5101	0.4913	0.4734	0.4563	0.4400	0.4244	0.4095	0.3954	0.3819	0.3690	0.3567	0.3450	0.3339	0.3232	0.3131
4	0.4560	0.4335	0.4123	0.3921	0.3730	0.3549	0.3378	0.3217	0.3066	0.2923	0.2738	0.2561	0.2541	0.2429	0.2322
5	0.4159	0.3904	0.3662	0.3435	0.3221	0.3022	0.2836	0.2662	0.2500	0.2350	0.2210	0.2080	0.1959	0.1846	0.1742
6	0.3847	0.3564	0.3297	0.3049	0.2818	0.2604	0.2407	0.2226	0.2059	0.1906	0.1766	0.1538	0.1521	0.1413	0.1315
7	0.3595	0.3285	0.2998	0.2731	0.2486	0.2263	0.2059	0.1874	0.1707	0.1556	0.1420	0.1297	0.1186	0.1086	0.0996
8	0.3387	0.3054	0.2746	0.2465	0.2209	0.1978	0.1771	0.1586	0.1421	0.1275	0.1145	0.1029	0.0927	0.0836	0.0756
9	0.3210	0.2856	0.2531	0.2236	0.1972	0.1737	0.1529	0.1347	0.1187	0.1047	0.0925	0.0819	0.0726	0.0645	0.0574
10	0.3058	0.2684	0.2343	0.2037	0.1767	0.1530	0.1324	0.1146	0.0993	0.0862	0.0749	0.0652	0.0569	0.0498	0.0437

ρ = 0.90, K = 0.75

YEAR	0	5	10	15	20	25	30	35	40	45	50	55	60	65	70
0	1.0000	1.0000	1.0000	1.0000	1.0000	1.0000	1.0000	1.0000	1.0000	1.0000	1.0000	1.0000	1.0000	1.0000	1.0000
1	0.7661	0.7587	0.7516	0.7446	0.7379	0.7314	0.7250	0.7189	0.7129	0.7071	0.7014	0.6959	0.6906	0.6854	0.6803
2	0.6522	0.6394	0.6270	0.6151	0.6035	0.5924	0.5816	0.5712	0.5611	0.5514	0.5420	0.5329	0.5241	0.5155	0.5073
3	0.5810	0.5637	0.5471	0.5311	0.5157	0.5009	0.4867	0.4731	0.4600	0.4475	0.4354	0.4239	0.4128	0.4022	0.3919
4	0.5308	0.5097	0.4894	0.4700	0.4514	0.4337	0.4168	0.4007	0.3853	0.3708	0.3569	0.3437	0.3312	0.3193	0.3080
5	0.4929	0.4683	0.4448	0.4223	0.4011	0.3809	0.3618	0.3438	0.3269	0.3109	0.2959	0.2318	0.2685	0.2560	0.2443
6	0.4628	0.4351	0.4086	0.3836	0.3600	0.3378	0.3170	0.2976	0.2795	0.2627	0.2470	0.2324	0.2189	0.2063	0.1947
7	0.4382	0.4076	0.3785	0.3511	0.3255	0.3016	0.2795	0.2591	0.2403	0.2230	0.2071	0.1925	0.1791	0.1668	0.1556
8	0.4176	0.3842	0.3526	0.3232	0.2958	0.2706	0.2475	0.2264	0.2073	0.1899	0.1741	0.1598	0.1469	0.1352	0.1245
9	0.3999	0.3639	0.3301	0.2988	0.2699	0.2437	0.2199	0.1985	0.1792	0.1620	0.1466	0.1329	0.1206	0.1096	0.0998
10	0.3846	0.3461	0.3102	0.2772	0.2471	0.2200	0.1958	0.1743	0.1552	0.1384	0.1236	0.1106	0.0991	0.0889	0.0800

*
*
* GROWTH
*

YEAR*	0	5	10	15	20	25	30	35	40	45	50	55	60	65	70
0	1.0000	1.0000	1.0000	1.0000	1.0000	1.0000	1.0000	1.0000	1.0000	1.0000	1.0000	1.0000	1.0000	1.0000	1.0000
1	0.8133	0.8072	0.8013	0.7955	0.7900	0.7845	0.7793	0.7741	0.7691	0.7643	0.7595	0.7549	0.7504	0.7460	0.7417
2	0.7179	0.7069	0.6963	0.6859	0.6759	0.6662	0.6568	0.6476	0.6388	0.6302	0.6218	0.6137	0.6058	0.5981	0.5907
3	0.6563	0.6411	0.6264	0.6121	0.5983	0.5850	0.5721	0.5596	0.5476	0.5359	0.5247	0.5139	0.5034	0.4933	0.4836
4	0.6118	0.5929	0.5745	0.5567	0.5396	0.5231	0.5072	0.4919	0.4773	0.4632	0.4497	0.4368	0.4244	0.4125	0.4012
5	0.5776	0.5552	0.5334	0.5124	0.4923	0.4730	0.4545	0.4359	0.4201	0.4041	0.3889	0.3744	0.3606	0.3475	0.3351
6	0.5501	0.5244	0.4995	0.4756	0.4527	0.4309	0.4102	0.3906	0.3721	0.3545	0.3380	0.3225	0.3078	0.2940	0.2810
7	0.5273	0.4985	0.4706	0.4440	0.4187	0.3947	0.3720	0.3508	0.3309	0.3122	0.2948	0.2786	0.2634	0.2493	0.2362
8	0.5079	0.4751	0.4455	0.4164	0.3888	0.3628	0.3386	0.3160	0.2950	0.2756	0.2577	0.2411	0.2258	0.2117	0.1987
9	0.4912	0.4566	0.4233	0.3918	0.3621	0.3345	0.3089	0.2852	0.2636	0.2437	0.2255	0.2090	0.1938	0.1800	0.1673
10	0.4765	0.4392	0.4034	0.3696	0.3381	0.3090	0.2823	0.2579	0.2358	0.2157	0.1976	0.1812	0.1664	0.1531	0.1410

P=0.90
K=0.80

YEAR*	0	5	10	15	20	25	30	35	40	45	50	55	60	65	70
0	1.0000	1.0000	1.0000	1.0000	1.0000	1.0000	1.0000	1.0000	1.0000	1.0000	1.0000	1.0000	1.0000	1.0000	1.0000
1	0.8603	0.8556	0.8510	0.8465	0.8422	0.8380	0.8339	0.8299	0.8260	0.8222	0.8185	0.8148	0.8113	0.8078	0.8044
2	0.7855	0.7768	0.7682	0.7599	0.7518	0.7439	0.7362	0.7288	0.7215	0.7144	0.7075	0.7007	0.6942	0.6878	0.6815
3	0.7358	0.7234	0.7113	0.6994	0.6879	0.6767	0.6658	0.6552	0.6449	0.6349	0.6252	0.6158	0.6066	0.5977	0.5891
4	0.6992	0.6833	0.6678	0.6527	0.6380	0.6238	0.6099	0.5965	0.5835	0.5709	0.5588	0.5470	0.5357	0.5247	0.5141
5	0.6705	0.6514	0.6327	0.6145	0.5968	0.5797	0.5631	0.5471	0.5317	0.5159	0.5026	0.4889	0.4758	0.4631	0.4510
6	0.6471	0.6249	0.6032	0.5820	0.5615	0.5417	0.5226	0.5043	0.4867	0.4699	0.4539	0.4385	0.4239	0.4100	0.3967
7	0.6275	0.6023	0.5775	0.5536	0.5304	0.5081	0.4867	0.4663	0.4468	0.4234	0.4108	0.3942	0.3785	0.3636	0.3495
8	0.6106	0.5825	0.5550	0.5283	0.5025	0.4779	0.4544	0.4321	0.4110	0.3912	0.3725	0.3549	0.3384	0.3228	0.3083
9	0.5959	0.5650	0.5347	0.5054	0.4772	0.4504	0.4250	0.4011	0.3786	0.3576	0.3380	0.3197	0.3027	0.2868	0.2720
10	0.5828	0.5492	0.5162	0.4844	0.4539	0.4251	0.3980	0.3727	0.3491	0.3272	0.3070	0.2882	0.2709	0.2549	0.2401

P=0.90
K=0.85

GROWTH

YEAR	0	5	10	15	20	25	30	35	40	45	50	55	60	65	70
0	1.0000	1.0000	1.0000	1.0000	1.0000	1.0000	1.0000	1.0000	1.0000	1.0000	1.0000	1.0000	1.0000	1.0000	1.0000
1	0.9070	0.9038	0.9007	0.8976	0.8947	0.8917	0.8889	0.8861	0.8834	0.8808	0.8782	0.8757	0.8732	0.8708	0.8684
2	0.8551	0.8489	0.8429	0.8369	0.8312	0.8255	0.8200	0.8146	0.8093	0.8041	0.7990	0.7941	0.7893	0.7845	0.7799
3	0.8197	0.8106	0.8018	0.7931	0.7846	0.7763	0.7682	0.7603	0.7525	0.7449	0.7375	0.7303	0.7232	0.7163	0.7096
4	0.7930	0.7813	0.7697	0.7584	0.7473	0.7364	0.7257	0.7154	0.7052	0.6953	0.6857	0.6763	0.6672	0.6583	0.6497
5	0.7717	0.7574	0.7432	0.7293	0.7156	0.7022	0.6891	0.6764	0.6640	0.6519	0.6402	0.6283	0.6178	0.6071	0.5968
6	0.7542	0.7373	0.7205	0.7040	0.6879	0.6720	0.6566	0.6416	0.6270	0.6129	0.5992	0.5860	0.5733	0.5610	0.5492
7	0.7392	0.7198	0.7006	0.6816	0.6629	0.6447	0.6270	0.6098	0.5932	0.5772	0.5613	0.5469	0.5327	0.5190	0.5059
8	0.7263	0.7044	0.6827	0.6612	0.6401	0.6196	0.5997	0.5804	0.5619	0.5442	0.5272	0.5109	0.4953	0.4805	0.4663
9	0.7149	0.6906	0.6664	0.6425	0.6190	0.5962	0.5742	0.5531	0.5328	0.5135	0.4950	0.4775	0.4608	0.4450	0.4300
10	0.7047	0.6780	0.6514	0.6250	0.5993	0.5743	0.5503	0.5274	0.5055	0.4847	0.4650	0.4464	0.4288	0.4122	0.3965

$\rho=0.90$, $K=0.90$

YEAR	0	5	10	15	20	25	30	35	40	45	50	55	60	65	70
0	1.0000	1.0000	1.0000	1.0000	1.0000	1.0000	1.0000	1.0000	1.0000	1.0000	1.0000	1.0000	1.0000	1.0000	1.0000
1	0.9536	0.9520	0.9504	0.9488	0.9472	0.9457	0.9443	0.9428	0.9414	0.9401	0.9387	0.9374	0.9361	0.9349	0.9336
2	0.9266	0.9234	0.9201	0.9170	0.9139	0.9109	0.9079	0.9050	0.9021	0.8993	0.8965	0.8938	0.8912	0.8886	0.8860
3	0.9077	0.9028	0.8980	0.8933	0.8886	0.8840	0.8795	0.8751	0.8707	0.8664	0.8622	0.8581	0.8541	0.8501	0.8462
4	0.8932	0.8868	0.8804	0.8740	0.8678	0.8616	0.8555	0.8495	0.8436	0.8379	0.8322	0.8266	0.8212	0.8158	0.8106
5	0.8815	0.8735	0.8655	0.8575	0.8497	0.8419	0.8342	0.8267	0.8193	0.8120	0.8048	0.7978	0.7910	0.7843	0.7778
6	0.8717	0.8621	0.8525	0.8430	0.8335	0.8241	0.8148	0.8057	0.7967	0.7879	0.7793	0.7709	0.7627	0.7547	0.7469
7	0.8632	0.8521	0.8409	0.8297	0.8186	0.8076	0.7967	0.7860	0.7755	0.7652	0.7552	0.7454	0.7359	0.7267	0.7177
8	0.8558	0.8432	0.8304	0.8176	0.8048	0.7921	0.7796	0.7673	0.7553	0.7436	0.7322	0.7211	0.7103	0.6999	0.6898
9	0.8492	0.8351	0.8207	0.8062	0.7918	0.7774	0.7633	0.7495	0.7360	0.7229	0.7101	0.6978	0.6858	0.6742	0.6630
10	0.8433	0.8277	0.8116	0.7955	0.7794	0.7634	0.7477	0.7323	0.7174	0.7029	0.6888	0.6753	0.6622	0.6496	0.6374

$\rho=0.90$, $K=0.95$

*
*
* GROWTH
*

YEAR*	0	5	10	15	20	25	30	35	40	45	50	55	60	65	70
0	1.0000	1.0000	1.0000	1.0000	1.0000	1.0000	1.0000	1.0000	1.0000	1.0000	1.0000	1.0000	1.0000	1.0000	1.0000
1	0.7092	0.7005	0.6920	0.6839	0.6760	0.6685	0.6611	0.6540	0.6471	0.6404	0.6339	0.6276	0.6215	0.6156	0.6098
2	0.5782	0.5639	0.5501	0.5369	0.5242	0.5121	0.5004	0.4891	0.4783	0.4679	0.4579	0.4483	0.4390	0.4301	0.4215
3	0.4997	0.4811	0.4634	0.4464	0.4303	0.4149	0.4002	0.3863	0.3730	0.3603	0.3482	0.3367	0.3258	0.3153	0.3054
4	0.4461	0.4240	0.4029	0.3830	0.3642	0.3464	0.3297	0.3139	0.2990	0.2849	0.2717	0.2593	0.2476	0.2366	0.2262
5	0.4065	0.3813	0.3575	0.3352	0.3143	0.2947	0.2764	0.2594	0.2436	0.2289	0.2152	0.2025	0.1907	0.1798	0.1696
6	0.3758	0.3478	0.3217	0.2973	0.2747	0.2538	0.2345	0.2168	0.2005	0.1856	0.1719	0.1594	0.1480	0.1375	0.1279
7	0.3510	0.3205	0.2923	0.2662	0.2423	0.2204	0.2005	0.1825	0.1661	0.1514	0.1381	0.1252	0.1154	0.1056	0.0969
8	0.3305	0.2978	0.2677	0.2401	0.2151	0.1926	0.1724	0.1543	0.1383	0.1240	0.1114	0.1001	0.0902	0.0814	0.0735
9	0.3131	0.2784	0.2466	0.2178	0.1920	0.1691	0.1488	0.1310	0.1155	0.1019	0.0900	0.0797	0.0706	0.0627	0.0558
10	0.2982	0.2616	0.2282	0.1984	0.1720	0.1489	0.1288	0.1115	0.0966	0.0838	0.0728	0.0634	0.0554	0.0484	0.0425

P=0.95 *K=0.70*

YEAR*	0	5	10	15	20	25	30	35	40	45	50	55	60	65	70
0	1.0000	1.0000	1.0000	1.0000	1.0000	1.0000	1.0000	1.0000	1.0000	1.0000	1.0000	1.0000	1.0000	1.0000	1.0000
1	0.7579	0.7504	0.7431	0.7361	0.7292	0.7226	0.7162	0.7100	0.7039	0.6981	0.6924	0.6868	0.6814	0.6761	0.6710
2	0.6428	0.6299	0.6175	0.6055	0.5940	0.5828	0.5721	0.5617	0.5517	0.5420	0.5325	0.5236	0.5148	0.5063	0.4981
3	0.5715	0.5543	0.5377	0.5218	0.5065	0.4918	0.4778	0.4643	0.4513	0.4389	0.4270	0.4156	0.4047	0.3942	0.3841
4	0.5215	0.5005	0.4804	0.4612	0.4428	0.4253	0.4086	0.3927	0.3776	0.3633	0.3496	0.3367	0.3244	0.3127	0.3016
5	0.4838	0.4595	0.4362	0.4141	0.3931	0.3732	0.3545	0.3368	0.3201	0.3044	0.2897	0.2758	0.2628	0.2505	0.2390
6	0.4541	0.4267	0.4006	0.3759	0.3527	0.3308	0.3104	0.2914	0.2736	0.2571	0.2417	0.2274	0.2142	0.2019	0.1904
7	0.4298	0.3995	0.3708	0.3439	0.3187	0.2953	0.2736	0.2536	0.2351	0.2182	0.2026	0.1883	0.1752	0.1632	0.1522
8	0.4094	0.3765	0.3454	0.3164	0.2896	0.2648	0.2422	0.2215	0.2028	0.1857	0.1703	0.1563	0.1436	0.1322	0.1218
9	0.3920	0.3565	0.3233	0.2925	0.2642	0.2384	0.2151	0.1941	0.1753	0.1535	0.1434	0.1299	0.1179	0.1072	0.0976
10	0.3763	0.3390	0.3037	0.2713	0.2418	0.2152	0.1915	0.1705	0.1518	0.1354	0.1209	0.1081	0.0969	0.0870	0.0782

P=0.95 *K=0.75*

* GROWTH

YEAR*	0	5	10	15	20	25	30	35	40	45	50	55	60	65	70
0	1.0000	1.0000	1.0000	1.0000	1.0000	1.0000	1.0000	1.0000	1.0000	1.0000	1.0000	1.0000	1.0000	1.0000	1.0000
1	0.3065	0.8003	0.7943	0.7884	0.7828	0.7773	0.7719	0.7667	0.7616	0.7567	0.7519	0.7472	0.7426	0.7382	0.7338
2	0.7098	0.6988	0.6880	0.6777	0.6676	0.6579	0.6484	0.6393	0.6304	0.6218	0.6135	0.6054	0.5975	0.5899	0.5824
3	0.6479	0.6327	0.6180	0.6038	0.5900	0.5767	0.5639	0.5515	0.5395	0.5280	0.5169	0.5061	0.4957	0.4857	0.4761
4	0.6035	0.5846	0.5663	0.5486	0.5316	0.5152	0.4995	0.4843	0.4698	0.4559	0.4426	0.4298	0.4176	0.4059	0.3946
5	0.5694	0.5471	0.5255	0.5047	0.4847	0.4656	0.4473	0.4299	0.4133	0.3975	0.3825	0.3682	0.3546	0.3418	0.3295
6	0.5421	0.5165	0.4918	0.4682	0.4456	0.4240	0.4036	0.3842	0.3659	0.3487	0.3324	0.3171	0.3026	0.2890	0.2763
7	0.5194	0.4908	0.4633	0.4369	0.4119	0.3882	0.3659	0.3450	0.3253	0.3070	0.2898	0.2739	0.2590	0.2451	0.2321
8	0.5002	0.4687	0.4385	0.4096	0.3824	0.3568	0.3329	0.3107	0.2900	0.2710	0.2533	0.2370	0.2220	0.2081	0.1953
9	0.4836	0.4494	0.4165	0.3854	0.3561	0.3289	0.3037	0.2804	0.2591	0.2396	0.2217	0.2054	0.1905	0.1769	0.1645
10	0.4691	0.4322	0.3968	0.3635	0.3325	0.3038	0.2775	0.2535	0.2317	0.2120	0.1942	0.1781	0.1636	0.1504	0.1386

$\rho = 0.95$
K=0.80

YEAR*	0	5	10	15	20	25	30	35	40	45	50	55	60	65	70
0	1.0000	1.0000	1.0000	1.0000	1.0000	1.0000	1.0000	1.0000	1.0000	1.0000	1.0000	1.0000	1.0000	1.0000	1.0000
1	0.8551	0.8502	0.8456	0.8410	0.8366	0.8323	0.8281	0.8241	0.8201	0.8162	0.8125	0.8088	0.8052	0.8016	0.7982
2	0.7791	0.7702	0.7616	0.7532	0.7451	0.7371	0.7294	0.7219	0.7146	0.7075	0.7006	0.6938	0.6872	0.6808	0.6746
3	0.7290	0.7165	0.7043	0.6925	0.6809	0.6697	0.6588	0.6483	0.6380	0.6280	0.6184	0.6090	0.5999	0.5910	0.5824
4	0.6923	0.6764	0.6609	0.6458	0.6311	0.6169	0.6031	0.5898	0.5769	0.5644	0.5523	0.5406	0.5294	0.5185	0.5080
5	0.6636	0.6445	0.6259	0.6077	0.5901	0.5731	0.5566	0.5407	0.5254	0.5107	0.4966	0.4830	0.4700	0.4575	0.4455
6	0.6402	0.6181	0.5964	0.5754	0.5550	0.5353	0.5164	0.4982	0.4809	0.4642	0.4483	0.4332	0.4187	0.4050	0.3918
7	0.6206	0.5955	0.5710	0.5472	0.5241	0.5020	0.4808	0.4606	0.4414	0.4231	0.4058	0.3894	0.3738	0.3591	0.3452
8	0.6038	0.5759	0.5485	0.5220	0.4965	0.4721	0.4489	0.4268	0.4060	0.3863	0.3678	0.3505	0.3341	0.3188	0.3044
9	0.5891	0.5584	0.5284	0.4993	0.4714	0.4449	0.4198	0.3961	0.3739	0.3532	0.3338	0.3157	0.2989	0.2832	0.2686
10	0.5762	0.5428	0.5101	0.4785	0.4484	0.4199	0.3931	0.3681	0.3448	0.3232	0.3031	0.2846	0.2675	0.2517	0.2370

$\rho = 0.95$
K=0.85

* GROWTH

YEAR*	0	5	10	15	20	25	30	35	40	45	50	55	60	65	70
0	1.0000	1.0000	1.0000	1.0000	1.0000	1.0000	1.0000	1.0000	1.0000	1.0000	1.0000	1.0000	1.0000	1.0000	1.0000
1	0.9035	0.9002	0.8970	0.8938	0.8908	0.8878	0.8849	0.8821	0.8793	0.8767	0.8740	0.8714	0.8689	0.8665	0.8641
2	0.8506	0.8443	0.8382	0.8322	0.8263	0.8206	0.8150	0.8096	0.8042	0.7990	0.7940	0.7890	0.7841	0.7794	0.7747
3	0.8147	0.8056	0.7967	0.7880	0.7795	0.7711	0.7630	0.7550	0.7473	0.7397	0.7323	0.7250	0.7180	0.7111	0.7044
4	0.7879	0.7761	0.7645	0.7532	0.7420	0.7311	0.7205	0.7101	0.7000	0.6901	0.6805	0.6712	0.6621	0.6533	0.6447
5	0.7665	0.7522	0.7380	0.7241	0.7104	0.6970	0.6840	0.6713	0.6589	0.6469	0.6352	0.6239	0.6130	0.6023	0.5920
6	0.7489	0.7320	0.7153	0.6988	0.6827	0.6669	0.6515	0.6366	0.6221	0.6080	0.5945	0.5814	0.5687	0.5555	0.5448
7	0.7340	0.7146	0.6954	0.6764	0.6578	0.6397	0.6221	0.6050	0.5885	0.5726	0.5573	0.5425	0.5284	0.5148	0.5018
8	0.7210	0.6992	0.6775	0.6561	0.6351	0.6147	0.5949	0.5758	0.5574	0.5398	0.5229	0.5068	0.4913	0.4766	0.4625
9	0.7096	0.6854	0.6613	0.6375	0.6142	0.5915	0.5696	0.5486	0.5285	0.5093	0.4910	0.4736	0.4571	0.4414	0.4264
10	0.6995	0.6729	0.6464	0.6201	0.5945	0.5697	0.5459	0.5231	0.5014	0.4808	0.4613	0.4423	0.4253	0.4089	0.3933

P=0.95
K=0.90

YEAR*	0	5	10	15	20	25	30	35	40	45	50	55	60	65	70
0	1.0000	1.0000	1.0000	1.0000	1.0000	1.0000	1.0000	1.0000	1.0000	1.0000	1.0000	1.0000	1.0000	1.0000	1.0000
1	0.9518	0.9501	0.9484	0.9468	0.9453	0.9437	0.9422	0.9408	0.9393	0.9379	0.9366	0.9352	0.9339	0.9326	0.9313
2	0.9242	0.9209	0.9176	0.9144	0.9113	0.9082	0.9052	0.9023	0.8994	0.8965	0.8938	0.8910	0.8884	0.8857	0.8832
3	0.9051	0.9001	0.8953	0.8905	0.8858	0.8812	0.8766	0.8721	0.8678	0.8635	0.8592	0.8551	0.8510	0.8471	0.8432
4	0.8904	0.8839	0.8775	0.8711	0.8648	0.8586	0.8525	0.8465	0.8406	0.8348	0.8291	0.8236	0.8181	0.8128	0.8076
5	0.8786	0.8705	0.8625	0.8545	0.8467	0.8389	0.8312	0.8236	0.8162	0.8089	0.8018	0.7948	0.7880	0.7813	0.7748
6	0.8687	0.8591	0.8495	0.8399	0.8304	0.8210	0.8117	0.8026	0.7937	0.7849	0.7763	0.7579	0.7598	0.7518	0.7440
7	0.8602	0.8491	0.8379	0.8267	0.8156	0.8045	0.7937	0.7830	0.7725	0.7623	0.7523	0.7425	0.7330	0.7238	0.7148
8	0.8528	0.8401	0.8274	0.8145	0.8017	0.7891	0.7766	0.7644	0.7524	0.7407	0.7293	0.7183	0.7075	0.6971	0.6870
9	0.8462	0.8320	0.8176	0.8032	0.7887	0.7744	0.7604	0.7466	0.7331	0.7200	0.7073	0.6950	0.6831	0.6715	0.6604
10	0.8403	0.8246	0.8086	0.7925	0.7764	0.7604	0.7448	0.7295	0.7146	0.7001	0.6861	0.6726	0.6596	0.6470	0.6349

P=0.95
K=0.95

* GROWTH

YEAR*	0	5	10	15	20	25	30	35	40	45	50	55	60	65	70
0	1.0000	1.0000	1.0000	1.0000	1.0000	1.0000	1.0000	1.0000	1.0000	1.0000	1.0000	1.0000	1.0000	1.0000	1.0000
1	0.7000	0.6912	0.6826	0.6744	0.6665	0.6588	0.6514	0.6443	0.6373	0.6306	0.6241	0.6177	0.6116	0.6056	0.5998
2	0.5682	0.5539	0.5401	0.5270	0.5144	0.5023	0.4906	0.4795	0.4687	0.4584	0.4435	0.4390	0.4298	0.4210	0.4125
3	0.4900	0.4715	0.4539	0.4371	0.4212	0.4060	0.3915	0.3777	0.3646	0.3521	0.3403	0.3290	0.3182	0.3079	0.2982
4	0.4368	0.4149	0.3942	0.3746	0.3560	0.3385	0.3220	0.3065	0.2919	0.2782	0.2652	0.2530	0.2416	0.2308	0.2206
5	0.3977	0.3729	0.3494	0.3275	0.3069	0.2877	0.2698	0.2531	0.2376	0.2232	0.2099	0.1975	0.1859	0.1752	0.1653
6	0.3674	0.3399	0.3142	0.2903	0.2681	0.2476	0.2287	0.2114	0.1955	0.1809	0.1676	0.1554	0.1442	0.1340	0.1247
7	0.3430	0.3131	0.2854	0.2598	0.2363	0.2149	0.1955	0.1779	0.1619	0.1476	0.1346	0.1229	0.1124	0.1029	0.0944
8	0.3228	0.2908	0.2612	0.2342	0.2098	0.1878	0.1680	0.1504	0.1348	0.1209	0.1085	0.0976	0.0879	0.0793	0.0716
9	0.3058	0.2718	0.2406	0.2124	0.1872	0.1648	0.1450	0.1277	0.1125	0.0992	0.0877	0.0776	0.0688	0.0611	0.0544
10	0.2912	0.2552	0.2226	0.1934	0.1677	0.1451	0.1256	0.1087	0.0941	0.0816	0.0710	0.0618	0.0539	0.0472	0.0414

ρ=1.00
K=0.70

YEAR*	0	5	10	15	20	25	30	35	40	45	50	55	60	65	70
0	1.0000	1.0000	1.0000	1.0000	1.0000	1.0000	1.0000	1.0000	1.0000	1.0000	1.0000	1.0000	1.0000	1.0000	1.0000
1	0.7500	0.7424	0.7350	0.7278	0.7209	0.7142	0.7077	0.7014	0.6953	0.6894	0.6837	0.6781	0.6726	0.6673	0.6622
2	0.6338	0.6209	0.6085	0.5965	0.5850	0.5738	0.5631	0.5527	0.5427	0.5331	0.5238	0.5148	0.5061	0.4977	0.4895
3	0.5625	0.5453	0.5288	0.5130	0.4979	0.4833	0.4694	0.4560	0.4432	0.4309	0.4192	0.4079	0.3971	0.3867	0.3768
4	0.5127	0.4919	0.4720	0.4529	0.4347	0.4174	0.4010	0.3853	0.3704	0.3563	0.3428	0.3301	0.3180	0.3065	0.2956
5	0.4754	0.4513	0.4283	0.4064	0.3857	0.3661	0.3476	0.3302	0.3138	0.2984	0.2839	0.2702	0.2575	0.2454	0.2341
6	0.4459	0.4188	0.3931	0.3687	0.3458	0.3243	0.3043	0.2855	0.2681	0.2519	0.2368	0.2228	0.2098	0.1977	0.1865
7	0.4219	0.3920	0.3637	0.3372	0.3124	0.2894	0.2681	0.2484	0.2303	0.2137	0.1984	0.1844	0.1716	0.1598	0.1490
8	0.4017	0.3693	0.3387	0.3102	0.2838	0.2595	0.2373	0.2170	0.1986	0.1819	0.1667	0.1530	0.1406	0.1294	0.1192
9	0.3846	0.3496	0.3169	0.2866	0.2588	0.2336	0.2107	0.1901	0.1717	0.1552	0.1404	0.1272	0.1155	0.1049	0.0955
10	0.3696	0.3324	0.2977	0.2658	0.2368	0.2108	0.1876	0.1669	0.1487	0.1326	0.1184	0.1059	0.0949	0.0851	0.0766

ρ=1.00
K=0.75

* GROWTH

ρ=1.00, K=0.80

YEAR	0	5	10	15	20	25	30	35	40	45	50	55	60	65	70
0	1.0000	1.0000	1.0000	1.0000	1.0000	1.0000	1.0000	1.0000	1.0000	1.0000	1.0000	1.0000	1.0000	1.0000	1.0000
1	0.8000	0.7937	0.7875	0.7816	0.7758	0.7702	0.7648	0.7595	0.7544	0.7494	0.7445	0.7398	0.7352	0.7307	0.7263
2	0.7021	0.6910	0.6802	0.6698	0.6597	0.6500	0.6405	0.6314	0.6225	0.6139	0.6056	0.5975	0.5896	0.5820	0.5746
3	0.6400	0.6248	0.6101	0.5959	0.5822	0.5689	0.5562	0.5438	0.5320	0.5205	0.5095	0.4988	0.4885	0.4786	0.4690
4	0.5956	0.5768	0.5586	0.5410	0.5241	0.5078	0.4922	0.4772	0.4628	0.4491	0.4359	0.4233	0.4112	0.3996	0.3885
5	0.5617	0.5395	0.5180	0.4974	0.4776	0.4587	0.4406	0.4234	0.4070	0.3914	0.3765	0.3624	0.3491	0.3364	0.3243
6	0.5345	0.5091	0.4847	0.4612	0.4388	0.4175	0.3973	0.3782	0.3602	0.3432	0.3271	0.3120	0.2978	0.2844	0.2718
7	0.5120	0.4836	0.4564	0.4303	0.4056	0.3822	0.3602	0.3395	0.3202	0.3021	0.2852	0.2695	0.2548	0.2411	0.2284
8	0.4929	0.4618	0.4318	0.4033	0.3764	0.3512	0.3276	0.3057	0.2854	0.2666	0.2492	0.2332	0.2184	0.2047	0.1921
9	0.4765	0.4426	0.4101	0.3794	0.3505	0.3237	0.2988	0.2759	0.2549	0.2357	0.2181	0.2020	0.1874	0.1740	0.1618
10	0.4621	0.4256	0.3907	0.3578	0.3272	0.2989	0.2730	0.2494	0.2280	0.2086	0.1910	0.1752	0.1609	0.1480	0.1363

ρ=1.00, K=0.85

YEAR	0	5	10	15	20	25	30	35	40	45	50	55	60	65	70
0	1.0000	1.0000	1.0000	1.0000	1.0000	1.0000	1.0000	1.0000	1.0000	1.0000	1.0000	1.0000	1.0000	1.0000	1.0000
1	0.8500	0.8451	0.8403	0.8357	0.8312	0.8268	0.8226	0.8185	0.8144	0.8105	0.8067	0.8029	0.7993	0.7957	0.7922
2	0.7729	0.7640	0.7553	0.7469	0.7387	0.7307	0.7229	0.7154	0.7080	0.7009	0.6940	0.6872	0.6806	0.6742	0.6680
3	0.7225	0.7100	0.6978	0.6859	0.6743	0.6631	0.6523	0.6417	0.6315	0.6215	0.6119	0.6025	0.5935	0.5847	0.5762
4	0.6857	0.6698	0.6543	0.6393	0.6246	0.6105	0.5967	0.5834	0.5706	0.5582	0.5462	0.5346	0.5235	0.5127	0.5023
5	0.6570	0.6379	0.6194	0.6013	0.5838	0.5668	0.5505	0.5347	0.5196	0.5050	0.4910	0.4775	0.4646	0.4522	0.4403
6	0.6337	0.6116	0.5901	0.5691	0.5489	0.5294	0.5106	0.4926	0.4753	0.4589	0.4431	0.4281	0.4138	0.4002	0.3872
7	0.6141	0.5892	0.5648	0.5411	0.5183	0.4963	0.4753	0.4553	0.4363	0.4182	0.4010	0.3848	0.3694	0.3549	0.3411
8	0.5974	0.5696	0.5425	0.5162	0.4909	0.4667	0.4437	0.4218	0.4012	0.3818	0.3635	0.3463	0.3302	0.3150	0.3008
9	0.5828	0.5523	0.5225	0.4937	0.4660	0.4397	0.4149	0.3915	0.3695	0.3490	0.3299	0.3120	0.2953	0.2798	0.2654
10	0.5699	0.5368	0.5043	0.4731	0.4432	0.4150	0.3885	0.3637	0.3407	0.3193	0.2995	0.2812	0.2643	0.2487	0.2342

* GROWTH

YEAR*	0	5	10	15	20	25	30	35	40	45	50	55	60	65	70
0	1.0000	1.0000	1.0000	1.0000	1.0000	1.0000	1.0000	1.0000	1.0000	1.0000	1.0000	1.0000	1.0000	1.0000	1.0000
1	0.9000	0.8956	0.8934	0.8902	0.8871	0.8840	0.8811	0.8782	0.8754	0.8727	0.8700	0.8674	0.8648	0.8623	0.8599
2	0.8462	0.8399	0.8337	0.8276	0.8217	0.8159	0.8103	0.8048	0.7995	0.7942	0.7891	0.7841	0.7792	0.7745	0.7698
3	0.8100	0.8009	0.7919	0.7831	0.7746	0.7662	0.7580	0.7501	0.7423	0.7347	0.7273	0.7201	0.7130	0.7061	0.6995
4	0.7830	0.7712	0.7595	0.7482	0.7371	0.7262	0.7155	0.7052	0.6951	0.6852	0.6757	0.6664	0.6573	0.6485	0.6399
5	0.7616	0.7472	0.7330	0.7191	0.7054	0.6921	0.6791	0.6664	0.6541	0.6421	0.6305	0.6193	0.6084	0.5978	0.5876
6	0.7439	0.7271	0.7104	0.6939	0.6778	0.6621	0.6468	0.6319	0.6175	0.6035	0.5900	0.5770	0.5644	0.5523	0.5406
7	0.7290	0.7096	0.6905	0.6716	0.6530	0.6350	0.6174	0.6005	0.5840	0.5682	0.5530	0.5384	0.5243	0.5109	0.4979
8	0.7161	0.6943	0.6727	0.6513	0.6305	0.6101	0.5904	0.5715	0.5532	0.5357	0.5189	0.5029	0.4875	0.4729	0.4589
9	0.7047	0.6805	0.6565	0.6328	0.6096	0.5871	0.5653	0.5444	0.5245	0.5054	0.4872	0.4700	0.4535	0.4379	0.4231
0	0.6946	0.6631	0.6415	0.6155	0.5901	0.5654	0.5417	0.5191	0.4975	0.4771	0.4577	0.4394	0.4220	0.4057	0.3902

P=1.00
K=0.90

YEAR*	0	5	10	15	20	25	30	35	40	45	50	55	60	65	70
0	1.0000	1.0000	1.0000	1.0000	1.0000	1.0000	1.0000	1.0000	1.0000	1.0000	1.0000	1.0000	1.0000	1.0000	1.0000
1	0.9500	0.9483	0.9466	0.9449	0.9433	0.9418	0.9402	0.9387	0.9373	0.9358	0.9344	0.9331	0.9317	0.9304	0.9291
2	0.9219	0.9185	0.9152	0.9120	0.9088	0.9057	0.9027	0.8997	0.8968	0.8939	0.8911	0.8883	0.8857	0.8830	0.8804
3	0.9025	0.8975	0.8926	0.8878	0.8831	0.8784	0.8738	0.8693	0.8649	0.8606	0.8564	0.8522	0.8482	0.8442	0.8403
4	0.8877	0.8812	0.8747	0.8683	0.8620	0.8558	0.8496	0.8436	0.8377	0.8319	0.8262	0.8207	0.8152	0.8099	0.8047
5	0.8758	0.8677	0.8597	0.8517	0.8438	0.8360	0.8283	0.8207	0.8133	0.8060	0.7989	0.7919	0.7851	0.7784	0.7719
6	0.8659	0.8563	0.8465	0.8370	0.8275	0.8181	0.8088	0.7997	0.7908	0.7820	0.7735	0.7551	0.7570	0.7490	0.7412
7	0.8574	0.8462	0.8350	0.8238	0.8127	0.8016	0.7908	0.7801	0.7697	0.7594	0.7495	0.7398	0.7303	0.7211	0.7122
8	0.8499	0.8373	0.8245	0.8116	0.7988	0.7862	0.7738	0.7615	0.7496	0.7379	0.7266	0.7156	0.7049	0.6945	0.6844
9	0.8433	0.8291	0.8147	0.8003	0.7859	0.7716	0.7575	0.7438	0.7304	0.7173	0.7047	0.6924	0.6805	0.6690	0.6579
10	0.8374	0.8217	0.8057	0.7896	0.7735	0.7576	0.7420	0.7267	0.7119	0.6975	0.6835	0.6701	0.6571	0.6445	0.6325

P=1.00
K=0.95

Table 5
Rate of Marginal Cost Decline

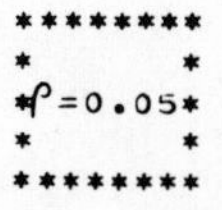

GROWTH 0/0 \ K 0/0	70	75	80	85	90	95
0.	2.479	2.005	1.558	1.137	0.739	0.360
5.	2.599	2.101	1.634	1.193	0.775	0.378
10.	2.717	2.198	1.709	1.247	0.811	0.395
15.	2.836	2.294	1.784	1.302	0.846	0.413
20.	2.954	2.389	1.858	1.357	0.882	0.430
25.	3.071	2.485	1.933	1.411	0.917	0.448
30.	3.189	2.580	2.007	1.466	0.953	0.465
35.	3.305	2.675	2.081	1.520	0.988	0.482
40.	3.422	2.769	2.155	1.574	1.023	0.499
45.	3.537	2.863	2.228	1.628	1.058	0.517
50.	3.653	2.957	2.301	1.681	1.093	0.534
55.	3.768	3.050	2.374	1.735	1.128	0.551
60.	3.883	3.144	2.447	1.788	1.163	0.568
65.	3.997	3.237	2.520	1.842	1.198	0.585
70.	4.111	3.329	2.592	1.895	1.232	0.602

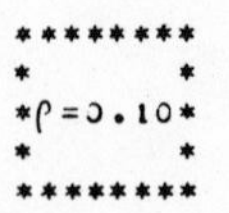

GROWTH 0/0 \ K 0/0	70	75	80	85	90	95
0.	4.786	3.878	3.022	2.210	1.438	0.703
5.	5.008	4.059	3.163	2.314	1.506	0.736
10.	5.228	4.239	3.304	2.417	1.574	0.769
15.	5.447	4.417	3.444	2.520	1.641	0.802
20.	5.665	4.595	3.583	2.622	1.708	0.835
25.	5.881	4.771	3.721	2.724	1.774	0.868
30.	6.095	4.946	3.858	2.825	1.841	0.900
35.	6.308	5.120	3.995	2.925	1.906	0.933
40.	6.520	5.293	4.130	3.025	1.972	0.965
45.	6.730	5.465	4.265	3.125	2.037	0.997
50.	6.939	5.636	4.400	3.224	2.102	1.029
55.	7.147	5.805	4.533	3.322	2.167	1.061
60.	7.353	5.974	4.666	3.420	2.231	1.092
65.	7.558	6.142	4.798	3.517	2.295	1.124
70.	7.761	6.308	4.929	3.614	2.358	1.155

ρ=0.15

GROWTH 0/0 \ K 0/0	70	75	80	85	90	95
0.	6.939	5.636	4.400	3.224	2.102	1.029
5.	7.250	5.890	4.599	3.371	2.199	1.077
10.	7.558	6.142	4.798	3.517	2.295	1.124
15.	7.862	6.391	4.994	3.662	2.390	1.171
20.	8.164	6.639	5.189	3.806	2.484	1.217
25.	8.463	6.884	5.382	3.949	2.578	1.264
30.	8.759	7.127	5.574	4.091	2.672	1.310
35.	9.053	7.368	5.764	4.231	2.764	1.355
40.	9.343	7.607	5.952	4.371	2.856	1.401
45.	9.631	7.843	6.139	4.509	2.947	1.446
50.	9.916	8.078	6.324	4.647	3.038	1.491
55.	10.198	8.310	6.508	4.783	3.128	1.535
60.	10.478	8.541	6.691	4.919	3.217	1.579
65.	10.756	8.770	6.872	5.053	3.306	1.623
70.	11.031	8.996	7.051	5.186	3.394	1.667

$\rho = 0.20$

GROWTH 0/0 \ K 0/0	70	75	80	85	90	95
0.	8.955	7.288	5.701	4.185	2.733	1.340
5.	9.343	7.607	5.952	4.371	2.856	1.401
10.	9.726	7.922	6.201	4.555	2.977	1.461
15.	10.105	8.233	6.447	4.738	3.098	1.520
20.	10.478	8.541	6.691	4.919	3.217	1.579
25.	10.848	8.845	6.932	5.097	3.335	1.638
30.	11.212	9.146	7.170	5.275	3.452	1.696
35.	11.573	9.444	7.406	5.450	3.568	1.753
40.	11.929	9.738	7.640	5.624	3.683	1.810
45.	12.281	10.029	7.871	5.796	3.797	1.867
50.	12.629	10.317	8.099	5.966	3.910	1.923
55.	12.973	10.602	8.326	6.135	4.021	1.978
60.	13.313	10.884	8.550	6.302	4.132	2.034
65.	13.649	11.162	8.772	6.468	4.242	2.088
70.	13.981	11.438	8.992	6.632	4.351	2.142

$\rho = 0.25$

GROWTH 0/0 \ K 0/0	70	75	80	85	90	95
0.	10.848	8.845	6.932	5.097	3.335	1.638
5.	11.303	9.221	7.229	5.319	3.481	1.710
10.	11.751	9.591	7.523	5.537	3.625	1.782
15.	12.193	9.957	7.813	5.753	3.768	1.853
20.	12.629	10.317	8.099	5.966	3.910	1.923
25.	13.058	10.673	8.382	6.177	4.049	1.992
30.	13.481	11.023	8.661	6.385	4.187	2.061
35.	13.898	11.369	8.937	6.591	4.324	2.129
40.	14.309	11.711	9.209	6.795	4.459	2.196
45.	14.715	12.048	9.478	6.996	4.593	2.263
50.	15.115	12.381	9.744	7.195	4.725	2.329
55.	15.509	12.709	10.006	7.391	4.856	2.394
60.	15.898	13.034	10.266	7.586	4.986	2.459
65.	16.282	13.354	10.522	7.778	5.114	2.523
70.	16.660	13.670	10.776	7.969	5.241	2.587

ρ=0.30

GROWTH 0/0 \ K 0/0	70	75	80	85	90	95
0.	12.629	10.317	8.099	5.966	3.910	1.923
5.	13.143	10.743	8.438	6.219	4.077	2.006
10.	13.649	11.162	8.772	6.468	4.242	2.088
15.	14.145	11.575	9.101	6.713	4.405	2.169
20.	14.634	11.981	9.425	6.956	4.566	2.250
25.	15.115	12.381	9.744	7.195	4.725	2.329
30.	15.587	12.774	10.059	7.430	4.882	2.407
35.	16.052	13.162	10.369	7.663	5.037	2.485
40.	16.510	13.544	10.675	7.893	5.191	2.561
45.	16.960	13.920	10.976	8.119	5.342	2.637
50.	17.403	14.291	11.274	8.343	5.491	2.712
55.	17.839	14.656	11.567	8.564	5.639	2.786
60.	18.269	15.016	11.857	8.782	5.785	2.859
65.	18.692	15.371	12.143	8.998	5.929	2.932
70.	19.108	15.721	12.425	9.210	6.072	3.004

$\rho = 0.35$

GROWTH 0/0 \ K 0/0	70	75	80	85	90	95
0.	14.309	11.711	9.209	6.795	4.459	2.196
5.	14.875	12.182	9.585	7.076	4.646	2.289
10.	15.430	12.644	9.954	7.352	4.830	2.381
15.	15.975	13.098	10.317	7.625	5.012	2.472
20.	16.510	13.544	10.675	7.893	5.191	2.561
25.	17.034	13.982	11.026	8.157	5.367	2.650
30.	17.549	14.413	11.372	8.417	5.541	2.737
35.	18.055	14.837	11.713	8.673	5.712	2.823
40.	18.551	15.254	12.048	8.926	5.881	2.908
45.	19.039	15.663	12.378	9.175	6.048	2.992
50.	19.519	16.066	12.703	9.421	6.213	3.075
55.	19.990	16.463	13.023	9.663	6.376	3.156
60.	20.453	16.853	13.338	9.901	6.536	3.237
65.	20.908	17.237	13.649	10.137	6.694	3.317
70.	21.356	17.615	13.955	10.369	6.851	3.396

$\rho = 0.40$

GROWTH 0/0 \ K 0/0	70	75	80	85	90	95
0.	15.898	13.034	10.266	7.586	4.986	2.459
5.	16.510	13.544	10.675	7.893	5.191	2.561
10.	17.108	14.044	11.076	8.194	5.392	2.662
15.	17.695	14.535	11.470	8.491	5.590	2.762
20.	18.269	15.016	11.857	8.782	5.785	2.859
25.	18.831	15.489	12.237	9.069	5.977	2.956
30.	19.383	15.952	12.610	9.351	6.166	3.051
35.	19.923	16.407	12.977	9.628	6.352	3.145
40.	20.453	16.853	13.338	9.901	6.536	3.237
45.	20.973	17.292	13.693	10.170	6.717	3.328
50.	21.483	17.722	14.042	10.434	6.895	3.418
55.	21.983	18.145	14.385	10.695	7.071	3.507
60.	22.474	18.561	14.722	10.951	7.244	3.595
65.	22.956	18.970	15.054	11.204	7.415	3.681
70.	23.429	19.372	15.381	11.453	7.583	3.766

GROWTH 0/0 \ K 0/0	70	75	80	85	90	95
0.	17.403	14.291	11.274	8.343	5.491	2.712
5.	18.055	14.837	11.713	8.673	5.712	2.823
10.	18.692	15.371	12.143	8.998	5.929	2.932
15.	19.314	15.894	12.564	9.316	6.143	3.039
20.	19.923	16.407	12.977	9.628	6.352	3.145
25.	20.519	16.908	13.383	9.935	6.559	3.249
30.	21.101	17.400	13.781	10.236	6.762	3.351
35.	21.671	17.882	14.171	10.533	6.961	3.452
40.	22.230	18.354	14.554	10.824	7.157	3.551
45.	22.776	18.817	14.930	11.110	7.351	3.649
50.	23.312	19.272	15.300	11.391	7.541	3.745
55.	23.837	19.718	15.663	11.668	7.728	3.840
60.	24.351	20.155	16.020	11.940	7.913	3.934
65.	24.855	20.585	16.370	12.208	8.095	4.026
70.	25.350	21.006	16.715	12.472	8.274	4.117

$\rho = 0.50$

GROWTH 0/0 \ K 0/0	70	75	80	85	90	95
0.	18.831	15.489	12.237	9.069	5.977	2.956
5.	19.519	16.066	12.703	9.421	6.213	3.075
10.	20.189	16.631	13.159	9.765	6.445	3.191
15.	20.844	17.183	13.605	10.103	6.672	3.306
20.	21.483	17.722	14.042	10.434	6.895	3.418
25.	22.107	18.250	14.470	10.759	7.114	3.529
30.	22.716	18.766	14.889	11.078	7.329	3.638
35.	23.312	19.272	15.300	11.391	7.541	3.745
40.	23.894	19.767	15.703	11.699	7.749	3.851
45.	24.464	20.251	16.098	12.000	7.953	3.954
50.	25.021	20.726	16.486	12.297	8.155	4.057
55.	25.566	21.191	16.866	12.588	8.352	4.157
60.	26.100	21.648	17.240	12.874	8.547	4.256
65.	26.623	22.095	17.607	13.155	8.739	4.354
70.	27.135	22.534	17.967	13.432	8.927	4.450

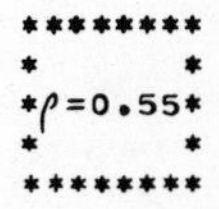

GROWTH 0/0 \ K 0/0	70	75	80	85	90	95
0.	20.189	16.631	13.159	9.765	6.445	3.191
5.	20.908	17.237	13.649	10.137	6.694	3.317
10.	21.609	17.829	14.128	10.500	6.939	3.441
15.	22.291	18.406	14.596	10.856	7.179	3.562
20.	22.956	18.970	15.054	11.204	7.415	3.681
25.	23.605	19.520	15.502	11.546	7.645	3.798
30.	24.238	20.059	15.941	11.880	7.872	3.913
35.	24.855	20.585	16.370	12.208	8.095	4.026
40.	25.458	21.099	16.791	12.530	8.313	4.137
45.	26.047	21.602	17.203	12.846	8.528	4.247
50.	26.623	22.095	17.607	13.155	8.739	4.354
55.	27.185	22.577	18.002	13.459	8.946	4.460
60.	27.735	23.049	18.390	13.758	9.150	4.564
65.	28.273	23.511	18.771	14.051	9.350	4.667
70.	28.800	23.964	19.145	14.339	9.547	4.767

$\rho=0.60$

GROWTH 0/0 \ K 0/0	70	75	80	85	90	95
0.	21.483	17.722	14.042	10.434	6.895	3.418
5.	22.230	18.354	14.554	10.824	7.157	3.551
10.	22.956	18.970	15.054	11.204	7.415	3.681
15.	23.663	19.570	15.543	11.576	7.666	3.809
20.	24.351	20.155	16.020	11.940	7.913	3.934
25.	25.021	20.726	16.486	12.297	8.155	4.057
30.	25.674	21.283	16.942	12.645	8.392	4.177
35.	26.311	21.828	17.387	12.987	8.624	4.296
40.	26.931	22.359	17.823	13.322	8.852	4.412
45.	27.537	22.878	18.250	13.650	9.076	4.526
50.	28.128	23.386	18.668	13.972	9.296	4.639
55.	28.705	23.883	19.077	14.287	9.511	4.749
60.	29.268	24.368	19.478	14.596	9.723	4.858
65.	29.819	24.844	19.871	14.900	9.931	4.965
70.	30.357	25.309	20.256	15.198	10.136	5.070

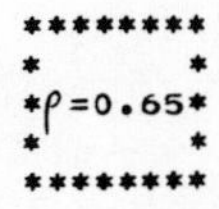

GROWTH 0/0 \ K 0/0	70	75	80	85	90	95
0.	22.716	18.766	14.889	11.078	7.329	3.638
5.	23.488	19.421	15.422	11.484	7.604	3.777
10.	24.238	20.059	15.941	11.880	7.872	3.913
15.	24.966	20.679	16.448	12.267	8.135	4.046
20.	25.674	21.283	16.942	12.646	8.392	4.177
25.	26.363	21.872	17.424	13.015	8.643	4.305
30.	27.033	22.446	17.895	13.377	8.890	4.431
35.	27.686	23.006	18.355	13.731	9.131	4.555
40.	28.322	23.553	18.805	14.077	9.368	4.676
45.	28.941	24.086	19.245	14.417	9.600	4.795
50.	29.545	24.607	19.675	14.749	9.828	4.912
55.	30.135	25.116	20.096	15.075	10.051	5.026
60.	30.710	25.614	20.508	15.394	10.270	5.139
65.	31.271	26.100	20.912	15.707	10.486	5.250
70.	31.819	26.576	21.307	16.014	10.697	5.359

$\rho = 0.70$

GROWTH 0/0 \ K 0/0	70	75	80	85	90	95
0.	23.894	19.767	15.703	11.699	7.749	3.851
5.	24.688	20.442	16.254	12.119	8.034	3.995
10.	25.458	21.099	16.791	12.530	8.313	4.137
15.	26.206	21.738	17.314	12.931	8.586	4.276
20.	26.931	22.359	17.823	13.322	8.852	4.412
25.	27.636	22.964	18.320	13.704	9.113	4.545
30.	28.322	23.553	18.805	14.077	9.368	4.676
35.	28.988	24.127	19.279	14.442	9.618	4.804
40.	29.637	24.686	19.741	14.799	9.862	4.929
45.	30.269	25.232	20.192	15.149	10.102	5.053
50.	30.884	25.765	20.633	15.491	10.337	5.173
55.	31.483	26.284	21.065	15.825	10.568	5.292
60.	32.068	26.792	21.487	16.153	10.794	5.409
65.	32.638	27.288	21.899	16.475	11.015	5.523
70.	33.194	27.772	22.303	16.790	11.233	5.636

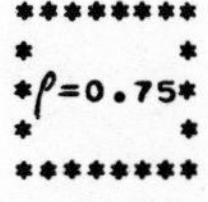

GROWTH 0/0 \ K 0/0	70	75	80	85	90	95
0.	25.021	20.726	16.486	12.297	8.155	4.057
5.	25.835	21.421	17.054	12.732	8.450	4.207
10.	26.623	22.095	17.607	13.155	8.739	4.354
15.	27.387	22.750	18.144	13.569	9.020	4.498
20.	28.128	23.386	18.668	13.972	9.296	4.639
25.	28.847	24.005	19.178	14.365	9.565	4.777
30.	29.545	24.607	19.675	14.749	9.828	4.912
35.	30.224	25.194	20.160	15.124	10.085	5.044
40.	30.884	25.765	20.633	15.491	10.337	5.173
45.	31.525	26.321	21.095	15.849	10.584	5.301
50.	32.150	26.864	21.546	16.200	10.826	5.425
55.	32.758	27.393	21.987	16.543	11.062	5.548
60.	33.350	27.909	22.417	16.878	11.294	5.668
65.	33.927	28.413	22.838	17.207	11.522	5.786
70.	34.490	28.905	23.250	17.529	11.745	5.901

$\rho=0.80$

GROWTH 0/0 \ K 0/0	70	75	80	85	90	95
0.	26.100	21.648	17.240	12.874	8.547	4.256
5.	26.931	22.359	17.823	13.322	8.852	4.412
10.	27.735	23.049	18.390	13.758	9.150	4.564
15.	28.514	23.718	18.942	14.183	9.440	4.713
20.	29.268	24.368	19.478	14.596	9.723	4.858
25.	30.000	25.000	20.000	15.000	10.000	5.000
30.	30.710	25.614	20.508	15.394	10.270	5.139
35.	31.399	26.211	21.004	15.778	10.535	5.275
40.	32.068	26.792	21.487	16.153	10.794	5.409
45.	32.718	27.358	21.958	16.520	11.047	5.539
50.	33.350	27.909	22.417	16.878	11.294	5.668
55.	33.965	28.446	22.866	17.229	11.537	5.793
60.	34.564	28.970	23.304	17.572	11.775	5.917
65.	35.147	29.481	23.733	17.907	12.008	6.038
70.	35.715	29.979	24.151	18.235	12.236	6.156

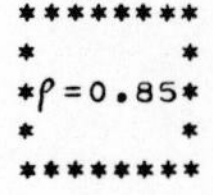

GROWTH 0/0 \ K 0/0	70	75	80	85	90	95
0.	27.135	22.534	17.967	13.432	8.927	4.450
5.	27.981	23.260	18.564	13.892	9.241	4.611
10.	28.800	23.964	19.145	14.339	9.547	4.767
15.	29.591	24.647	19.708	14.774	9.845	4.920
20.	30.357	25.309	20.256	15.198	10.136	5.070
25.	31.100	25.952	20.789	15.611	10.420	5.216
30.	31.819	26.576	21.307	16.014	10.697	5.359
35.	32.517	27.183	21.812	16.406	10.968	5.499
40.	33.194	27.772	22.303	16.790	11.233	5.636
45.	33.851	28.346	22.783	17.164	11.492	5.770
50.	34.490	28.905	23.250	17.529	11.745	5.901
55.	35.111	29.449	23.706	17.886	11.993	6.030
60.	35.715	29.979	24.151	18.235	12.236	6.156
65.	36.303	30.496	24.586	18.577	12.474	6.280
70.	36.875	31.000	25.010	18.911	12.707	6.402

ρ=0.90

GROWTH O/O \ K O/O	70	75	80	85	90	95
0.	28.128	23.386	18.668	13.972	9.296	4.639
5.	28.988	24.127	19.279	14.442	9.618	4.804
10.	29.819	24.844	19.871	14.900	9.931	4.965
15.	30.622	25.538	20.446	15.345	10.237	5.122
20.	31.399	26.211	21.004	15.778	10.535	5.275
25.	32.150	26.864	21.546	16.200	10.826	5.425
30.	32.878	27.497	22.074	16.610	11.109	5.572
35.	33.583	28.112	22.587	17.011	11.386	5.715
40.	34.267	28.709	23.086	17.401	11.657	5.855
45.	34.930	29.290	23.573	17.782	11.921	5.993
50.	35.574	29.856	24.047	18.154	12.179	6.127
55.	36.200	30.406	24.510	18.517	12.432	6.259
60.	36.808	30.941	24.961	18.872	12.680	6.388
65.	37.400	31.463	25.401	19.219	12.922	6.514
70.	37.975	31.972	25.831	19.558	13.159	6.638

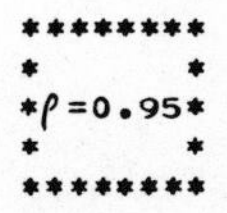

GROWTH 0/0 \ K 0/0	70	75	80	85	90	95
0.	29.082	24.208	19.345	14.494	9.653	4.822
5.	29.955	24.961	19.968	14.975	9.983	4.991
10.	30.797	25.689	20.571	15.442	10.304	5.156
15.	31.610	26.394	21.156	15.896	10.616	5.317
20.	32.395	27.077	21.724	16.338	10.921	5.474
25.	33.155	27.738	22.275	16.767	11.218	5.628
30.	33.889	28.380	22.811	17.185	11.507	5.778
35.	34.601	29.002	23.331	17.593	11.789	5.924
40.	35.290	29.606	23.838	17.990	12.065	6.068
45.	35.959	30.193	24.331	18.377	12.335	6.208
50.	36.608	30.764	24.812	18.755	12.598	6.345
55.	37.237	31.319	25.280	19.123	12.855	6.479
60.	37.849	31.860	25.736	19.483	13.107	6.611
65.	38.443	32.386	26.181	19.835	13.353	6.740
70.	39.021	32.898	26.616	20.179	13.594	6.866

ρ=1.00

GROWTH 0/0 \ K 0/0	70	75	80	85	90	95
0.	30.000	25.000	20.000	15.000	10.000	5.000
5.	30.884	25.765	20.633	15.491	10.337	5.173
10.	31.736	26.503	21.247	15.967	10.665	5.342
15.	32.557	27.218	21.841	16.429	10.984	5.507
20.	33.350	27.909	22.417	16.878	11.294	5.668
25.	34.117	28.578	22.977	17.315	11.597	5.824
30.	34.857	29.227	23.520	17.740	11.892	5.977
35.	35.574	29.856	24.047	18.154	12.179	6.127
40.	36.269	30.466	24.560	18.557	12.460	6.273
45.	36.941	31.058	25.059	18.950	12.734	6.416
50.	37.593	31.634	25.545	19.333	13.001	6.556
55.	38.226	32.194	26.018	19.706	13.263	6.693
60.	33.840	32.738	26.479	20.071	13.519	6.827
65.	39.437	33.268	26.929	20.427	13.769	6.958
70.	40.016	33.783	27.367	20.775	14.013	7.086

Table 6
Future Spending

GROWTH O/O \ K O/O	70	75	80	85	90	95
0	1.00000	1.00000	1.00000	1.00000	1.00000	1.00000
5	1.02397	1.02895	1.03364	1.03806	1.04224	1.04622
10	1.04735	1.05734	1.06676	1.07569	1.08418	1.09227
15	1.07020	1.08519	1.09940	1.11293	1.12583	1.13817
20	1.09254	1.11255	1.13159	1.14978	1.16720	1.18392
25	1.11440	1.13943	1.16335	1.18628	1.20831	1.22953
30	1.13582	1.16588	1.19471	1.22244	1.24918	1.27500
35	1.15682	1.19190	1.22568	1.25827	1.28980	1.32035
40	1.17743	1.21753	1.25628	1.29380	1.33020	1.36557
45	1.19766	1.24273	1.28653	1.32902	1.37038	1.41067
50	1.21753	1.26767	1.31644	1.36397	1.41034	1.45566
55	1.23706	1.29222	1.34604	1.39864	1.45011	1.50054
60	1.25628	1.31644	1.37533	1.43305	1.48968	1.54531
65	1.27518	1.34035	1.40433	1.46721	1.52906	1.58997
70	1.29380	1.36397	1.43305	1.50112	1.56827	1.63454
75	1.31213	1.38729	1.46149	1.53481	1.60730	1.67901
80	1.33020	1.41034	1.48968	1.56827	1.64615	1.72338
85	1.34801	1.43313	1.51762	1.60151	1.68485	1.76767
90	1.36557	1.45566	1.54531	1.63454	1.72338	1.81186
95	1.38290	1.47795	1.57277	1.66737	1.76177	1.85597
100	1.40000	1.50000	1.60000	1.70000	1.80000	1.90000

Table 7
Average Cost

GROWTH 0/0 \ K 0/0	70	75	80	85	90	95
0	1.00000	1.00000	1.00000	1.00000	1.00000	1.00000
5	0.97521	0.97995	0.98442	0.98863	0.99261	0.99640
10	0.95214	0.96122	0.96978	0.97790	0.98562	0.99297
15	0.93061	0.94364	0.95600	0.96776	0.97898	0.98971
20	0.91045	0.92712	0.94299	0.95815	0.97267	0.98660
25	0.89152	0.91155	0.93068	0.94903	0.96665	0.98362
30	0.87371	0.89683	0.91901	0.94034	0.96090	0.98077
35	0.85691	0.88289	0.90791	0.93205	0.95541	0.97804
40	0.84102	0.86966	0.89734	0.92414	0.95014	0.97541
45	0.82597	0.85709	0.88726	0.91657	0.94509	0.97288
50	0.81169	0.84511	0.87763	0.90931	0.94023	0.97044
55	0.79811	0.83369	0.86841	0.90235	0.93555	0.96809
60	0.78517	0.82278	0.85958	0.89566	0.93105	0.96582
65	0.77284	0.81234	0.85111	0.88922	0.92671	0.96362
70	0.76106	0.80233	0.84297	0.88301	0.92251	0.96149
75	0.74979	0.79274	0.83514	0.87703	0.91845	0.95943
80	0.73900	0.78352	0.82760	0.87126	0.91453	0.95744
85	0.72865	0.77466	0.82033	0.86568	0.91073	0.95550
90	0.71872	0.76614	0.81332	0.86028	0.90704	0.95361
95	0.70918	0.75792	0.80655	0.85506	0.90347	0.95178
100	0.70000	0.75000	0.80000	0.85000	0.90000	0.95000

References

References

[1] The Boston Consulting Group, *Perspectives on Experience*, Boston: The Boston Consulting Group, 1968.

[2] Mannis, R.D., *Eastern Pharmaceutical Manufacturers Research Group Meeting*, CIBA, 1968.

Catry, B., and Chevalier, M., "Market Share Strategy and the Product Life Cycle," *Journal of Marketing*, vol. 38, October 1974, pp. 29-34.

Chevalier, M., "The Strategy Spectre behind Your Market Share," *European Business*, Summer 1972.

[3] Zakon, A.J., and Henderson, B.D., "Financial Myopia," *California Management Review*, vol. xi, no. 2, 1968.

[4] Vielliard, E., "La croissance des entreprises en franchising," (unpublished thesis, 1974, Faculty of Administration, University of Sherbrooke, Québec).

[5] Alberts, W.W., and Archer, S.H., "Some Evidence on the Effect of Company Size on the Cost of Equity Capital," *Journal of Financial and Quantitative Analysis*, vol. 8, March 1973.

Archer, S.H., and Faerber, L.G., "Firm Size and the Cost of Equity Capital," *Journal of Finance*, vol. 21, March 1966.

Brigham, E.F., and Smith, K.V., "The Cost of Capital to the Small Firm," *Engineering Economist*, vol. 13, Fall 1967.

[6] Abernathy, William J., and Wayne, Kenneth, "Limits of the Learning Curve," *Harvard Business Review*, Sept./Oct. 1974.

[7] Sallenave, J.P., *La stratégie de l'entreprise face à la concurrence*, Editions d'Organisation, Paris, 1973.

[8] Bienaymé, A., *La croissance des entreprises—Tome II: Analyse dynamique de la concurrence industrielle*, p. 227, Ed. Bordas, Paris, 1973.

[9] Conley, P., "Experience Curves as a Planning Tool," *IEEE Spectrum*, June 1970.

Index

About the Author

Jean-Paul Sallenave is professor of marketing at the University of Sherbrooke, Québec. He also teaches Business Policy and Corporate Strategy in the graduate programs of the University of Sherbrooke. He has been Visiting Professor at the Institut National du Marketing, France, and at the Graduate School of Business of Columbia University.

Dr. Sallenave attended the Sorbonne and the Ecole Supérieure des Sciences Economiques et Commerciales in Paris, France, before graduating from the MBA program at Stanford University, California, where he was a Fulbright scholar. He received a doctorate from the University of Aix-Marseille, France, concentrating in competitive dynamics and strategy.

In his professional business career, Dr. Sallenave worked as a staff consultant with the Boston Consulting Group, Boston, specializing in corporate strategy. He then founded Eurac inc., a consulting firm specialized in international development and franchising.

Dr. Sallenave is the author of several articles on the subject of corporate strategy, institutional marketing, and franchising. His book *La Stratégie de l'entreprise face à la concurrence*, Ed. d'Organisation, Paris, 1973, won the Management Book-of-the-Year award in France in 1975.

Dr. Sallenave is an officer of the Canadian Association of Administrative Sciences and the Awards Chairman for the Samuel Bronfman Foundation.